BAD FRIDAY

The Great and Terrible 1964 Alaska Earthquake

LEW FREEDMAN

Epicenter Press

Epicenter is a regional press publishing nonfiction books about the arts, history, environment, and diverse cultures and lifestyles of Alaska and the Pacific Northwest.

Acquisitions Editor: Lael Morgan
Photo editor: Dan Langager
Cover & text design: Victoria Michael

PHOTO CREDITS
Front cover--View of Fourth Avenue, Anchorage, March 1964, U.S. Geological Survey Photographic Library (photo colorized by Tom Goodson); back cover--A home lies in ruins in the Turnagain neighborhood of Anchorage, Chuck Lastufka.
Text--page 125, Fourth Avenue, Anchorage, USGS; page 126-127, JC Penney store, Anchorage, John Schandelmeier; page 128-129, earthquake graben, downtown Anchorage, USGS; page 130, ruined homes, Turnagain area, Chuck Lastufka; page 131, ruined home, Turnagain area, Alaska Earthquake Archives Committee Collection, Accession Number UAF-1972-153-79, Archives, Alaska and Polar Regions Collections, Rasmuson Library, University of Alaska Fairbanks; page 132, rescue workers comb through demolished homes in Turnagain area, Alaska Earthquake Archives Committee Collection, #UAF-1972-153-70, Archives, Alaska and Polar Regions Collections, Rasmuson Library, UA Fairbanks; page 133 (top), Governor Egan surveys damage, William A. Egan Papers, #UAF-1985-120-286, Archives, Alaska and Polar Regions Collections, Rasmuson Library, UA Fairbanks; page 133 (bottom), cleanup begins on Fourth Avenue in Anchorage, USGS; pages 134-135, aerial of landslide area, Turnagain Heights, USGS; pages 136-137, sagging bluff near Alaska Native Medical Center, USGS: page 138, abandoned car on main street in Valdez, Alaska Earthquake Archives Committee Collection, #UAF-1972-153-276, Archives, Alaska and Polar Regions Collections, Rasmuson Library, UA Fairbanks; page 139, oil tanks burning in Valdez, National Oceanic and Atmospheric Administration; page 140, Alaska Railroad damage, USGS; page 141, Seward Highway damage, USGS; pages 142-143, cleaning up debris in Valdez, Alaska Earthquake Archives Committee Collection, #UAF-1972-153-296, Archives, Alaska and Polar Regions Collections, Rasmuson Library, UA Fairbanks; pages 144-146, Kodiak waterfront, NOAA; page 147, Valdez scenes, NOAA; pages 148-149, aerial of old and new Valdez, USGS.

Library of Congress Control Number: 2013933972

ISBN 978-1-935347-24-8
10 9 8 7 6 5 4 3 2
Printed in Canada

Thanks to everyone who revisited in interviews the dramatic and painful memories of the 1964 Great Alaska Earthquake and the days that followed.

"It was an awful sensation--like nothing I have experienced before. I knew it was an earthquake when it started, but never imagined one so horrible. The entire house seemed to be bending and twisting, back and forth. I kept thinking it had to stop, it could not possibly get worse, but it kept getting worse. I never have felt so near to death."

—Edna Kruckewitt, Anchorage, Alaska

Table of Contents

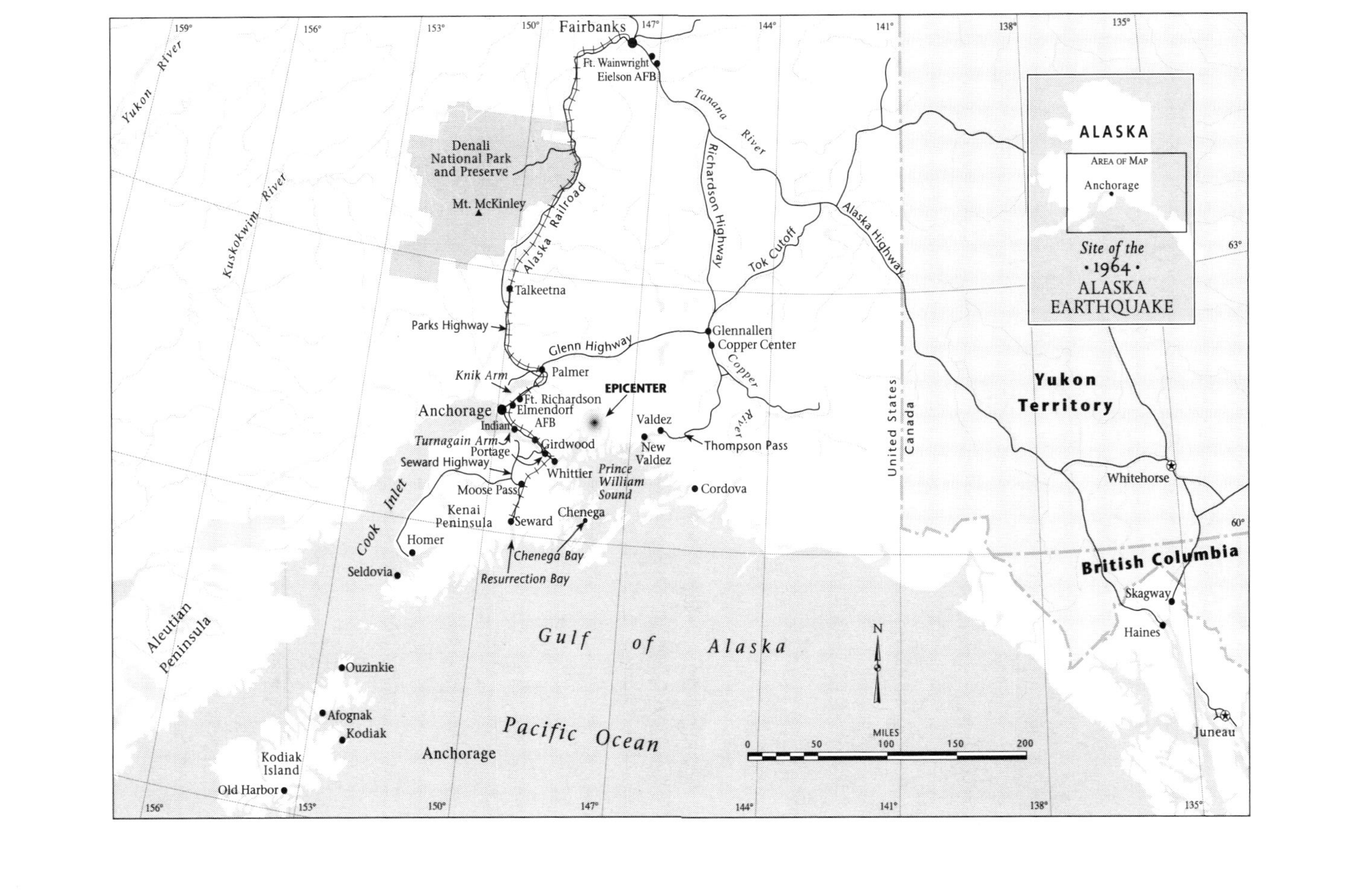
ALASKA
AREA OF MAP
Anchorage
Site of the
• 1964 •
ALASKA
EARTHQUAKE
Fairbanks
Ft. Wainwright
Eielson AFB
Tanana River
Richardson Highway
Tok Cutoff
Alaska Highway
Yukon River
Kuskokwim River
Denali National Park and Preserve
Mt. McKinley
Alaska Railroad
Talkeetna
Parks Highway
Glenn Highway
Glennallen
Copper Center
Copper River
Palmer
Knik Arm
EPICENTER
Ft. Richardson
Anchorage
Elmendorf AFB
Indian
Turnagain Arm
Portage
Girdwood
Valdez
New Valdez
Thompson Pass
Seward Highway
Whittier
Prince William Sound
Moose Pass
Cordova
Kenai Peninsula
Seward
Chenega
Cook Inlet
Homer
Chenega Bay
Resurrection Bay
Seldovia
United States
Canada
Yukon Territory
Whitehorse
British Columbia
Skagway
Haines
Juneau
Aleutian Peninsula
Gulf of Alaska
Pacific Ocean
Anchorage
Ouzinkie
Afognak
Kodiak
Kodiak Island
Old Harbor
N
MILES
0 50 100 150 200
159° 156° 153° 150° 147° 144° 141° 138° 135°
63° 60°

Facts and figures about the 1964 Good Friday Earthquake

Date: March 27, 1964

Magnitude: 9.2 on Richter scale

Area in which quake was felt: 500,000 square miles

Duration of quake: Maximum of 5 minutes 30 seconds in some places

Aftershocks: Ten quakes of a magnitude 6.0 or greater were felt within 24 hours.

Rank in world history: Second most-powerful earthquake ever recorded. The most powerful quake occurred in Chile registering 9.5 on the Richter scale on May 22, 1960.

Epicenter: Beneath Prince William Sound, 78 miles east of Anchorage

Fatalities: 131

Fatalities beyond Alaska: 16

Estimated property damage (Alaska): $311 million

Recovery aid: $330 million from government and private funds

Hardest-hit areas: Severe property damage and loss of life sustained in downtown Anchorage and in the city's Turnagain neighborhood; in the towns of Seward, Valdez, and Kodiak; and in several villages on Prince William Sound and the Gulf of Alaska.

Communities slammed by powerful tsunamis: Kodiak and Kodiak Island coastal villages, Seward, Chenega Bay, Whittier, and Valdez in Alaska; Prince Rupert, British Columbia; Crescent City, California; Beverly Beach State Park, Oregon.

Introduction

The big shake began at 5:36 p.m. on March 27, 1964, the Good Friday prelude to an Easter Sunday holiday gone bad.

There was no school that day in Anchorage, Alaska, the end of the nine-to-five workday for many, a day off for others. There were people sitting in bars who wouldn't have been if not for the end of the work week. There were children who had gone to work with their parents. Housewives were home cooking supper.

The time of day and the type of day, a quasi-holiday, played a role in the fate of where thousands of people were when the earth split open, the terra firma under their feet began shaking violently, and the second-largest earthquake in history created a frightening rumble.

Death, destruction, and fear were early consequences of what later was measured as a 9.2-magnitude earthquake. For five and a half minutes the ground vibrated, shaking houses off of their foundations, cracking open downtown streets, whipping trees back and forth so hard that their tops swayed over sideways, touching the ground. For those who withstood the shaking and power of the quake, those minutes seemed like an eternity.

People held onto solid buildings that could withstand the shock. People ran for cover. People hugged one another. The scene was similar throughout south-central Alaska when the Prince William Sound earthquake ripped through neighborhoods, uprooted trees, smashed previously smooth asphalt roads into rubble, ruptured gas lines, instigated small fires, and spawned a tsunami that killed people more than 1,800 miles to the south in California.

The towns of Valdez and Seward were nearly destroyed. The village of Chenega Bay was wiped off the map. It was somewhat miraculous that the death toll was only 115 in Alaska and 131 in all. But Alaska was young then, only five years into statehood, and more sparsely populated. An earthquake of similar strength today, affecting the same places, would produce many more casualties, much more damage, and given the comparative sophistication in communications technology, almost certainly would receive one hundred times the worldwide attention.

In the decades following the Great Alaska Earthquake, the earth's tectonic plates have crushed against one another many times in dramatic and violent ways. Earthquakes we are most familiar with have disrupted a World Series, killing Americans in the San Francisco Bay Area; devastated Haiti; torn asunder Japan, and in some instances generated the most awesome and fright-inducing tsunamis ever recorded.

Yet as cataclysmic as those earthquakes have been, none has matched the Alaska earthquake's sheer measured power. Much of Americans' earthquake knowledge, or focus, is fixated on California. One day, people sarcastically say, there will be an earthquake so strong that the nation's most populous state will break off and float away into the Pacific Ocean.

However, Alaska has more frequent and stronger earthquakes than California on a regular basis. Alaska is in greater peril. California is far more densely populated, so the death toll there would likely be higher, but no one wishes to imagine an extremely powerful earthquake hitting Alaska in the middle of winter, cutting off communication and heat.

Among Alaskans who survived the Good Friday Earthquake, memories remained clear and the experience vivid a half-century later.

So does the fear. Whenever an earthquake strikes Alaska—and they are very common—those who lived through the Big One begin counting seconds. They remain on edge until the overhead light fixture and the shaking bookcase stop vibrating. A five-second earthquake is barely noticeable. A thirty-second earthquake is interminable. A five-and-a-half-minute earthquake is unimaginable.

Time measured in seconds is how 1964 earthquake survivors have come to measure the severity of an earthquake today. Their relief is palpable when a quake, no matter how powerful or mild the shaking, ends within seconds. That is how they know the Big One has not struck Alaska for the second time in their lives. Once was enough.

1

The Nerland Family

Anchorage woke up innocent and went to bed scared and scarred.

With a population of about 82,000, Alaska's largest city was still somewhat of a frontier town in the early 1960s. The former U.S. territory bought from Russia for $7.2 million in 1867 was still feeling the glow from becoming a state in 1959.

Alaska had a boomer mentality. It was a resource-rich state with oil being pumped on the Kenai Peninsula and more oil sure to be found north of the Arctic Circle. Good times were ahead.

Sure, Anchorage was 1,500 miles from Seattle, and, yes, it still took awhile to get there even in increasingly larger and faster jet planes, and, yes, the cost of those long-distance telephone calls could add up. But Alaska was a place that attracted hardy people who wanted to start new lives, who saw it as an outdoorsman's paradise and a businessman's potential gold mine.

Except for the Native peoples who had been in the territory for an estimated ten thousand years, Alaska had been settled by men with a glint in their eye and fortune on their minds. The northern gold rush, beginning in 1898, brought the multitudes to the Klondike and later to Nome and Fairbanks.

World War II jump-started a different type of boom. Alaska was strategically placed for war against Japan, and the Japanese brought the war to Alaska's shores by invading the Aleutian Islands. Air Force bases, such as Elmendorf in Anchorage and Eielson near Fairbanks, and Army posts, such as Fort Richardson in Anchorage and Fort Wainwright in Fairbanks, transformed Alaska's two biggest cities into bustling places during the war years. The military built the Alaska Highway through Canada to move troops and supplies north.

The military remained after the war. Year after year young soldiers rotated through Alaska, learned about the state, and mustered out, with some staying in the state. During the cold war, Alaska became the first line of defense against the Soviet Union, the one-time World War II ally.

In those immediate years after statehood was approved, Alaska was finding its new identity, but the can-do spirit of the pioneers still prevailed.

All such resolve and optimism was about to be tested by this unforeseen natural disaster. Areas in Alaska and in the seas nearby are geologically volatile, and even today, earthquake forecasting is an inexact science. Once in a while scientists warn of seismic danger brewing, but not so in 1964.

That's what it was like that day when the earth got the tremors, almost like a potpourri of dirt, rock, snow, and grass had been poured into a mixer and blended into a chemically made bomb. No one saw it coming.

IT WAS A PEACEFUL day in Anchorage, a little quieter than normal because of the school holiday. Still, the stores were open and customers were out shopping at places like the new JC Penney store on Fourth Avenue downtown.

The workday at Nerland's Home Furnishings was out of the ordinary because, with no school, Steve Nerland and his brother Rick went to work in the family furniture store after the family attended services at the Episcopal Church.

The Nerlands were a fourth-generation Alaska family. Steve and Rick's great-grandfather Andrew came north from Seattle in 1898, chasing the dream of making it rich, if not from the gold, then from the prospectors who panned for gold.

Andrew Nerland established his business in Dawson City, Yukon Territory. When the gold ran out and people left, he moved on to Fairbanks in 1904 after gold was discovered there. Andrew Nerland began life in Norway and at age eighteen migrated to Seattle. He became partners with Anderson Brothers, painters and paper hangers, but was wooed to the gold territory by the first boatload of gold unloaded at the docks in Seattle.

Nerland sought to stake a Klondike claim but was too late, with much of the best land already claimed, so instead he opened a store in Dawson to serve the miners. He also wrote articles for the *Washington Posten*, a Norwegian newspaper. He probably applied the first wallpaper in Dawson.

"They used cheesecloth to stretch over the logs, and then they sized it and tightened it up, looking almost like sheetrock, except it wasn't solid," said Jerry Nerland, Andrew's grandson and Steve and Rick's father. "They used a balloon ceiling, they called it, stretching this fabric across the room, and tacked it on either side of the wall. But that became quite a fire hazard when they had chimneys going through it."

AFTER DEPARTING DAWSON and settling in Fairbanks, Andrew Nerland bought out Anderson Brothers, but then moved back to Seattle. In 1930, Jerry Nerland and his parents returned to Fairbanks, where they opened a store, and in 1962 the Nerlands established a second store on Fifth Avenue in downtown Anchorage. At the time of the quake, Steve Nerland was enrolled in his first year at the new Central Junior High School. He and brother Rick did odd jobs around the store.

When the first tremors were felt, Steve, then thirteen, happened to be standing next to a six-foot post in the store's office where a clerk and receptionist worked. When the building began to shake, he grabbed the post with two hands and squeezed tight.

"I was just hugging it, hanging on to it," he said.

A woman who worked in the store grasped the same post, too. Many chandeliers hanging by chains from the ceiling began to move. Then the power went. Through natural light coming in from the front windows, the Nerlands were stunned to see glass chandeliers swinging back and forth so violently that they crashed into the ceiling and shattered. Steve's mother, Maxine, and a receptionist began screaming.

It was chaos in the store. Roger, a younger brother, was watching television in the basement where an electronics salesman grabbed him and shoved him under a table. Brian, the fourth and youngest son, was home with a babysitter.

Rick Nerland, who was eleven, was sent on an errand a few minutes before the earth began shaking. While his father and brother recalled that he was taking mail to the post office, Rick said he actually was on his way to send telegrams at the Alaska Communications System office in the federal building one street over on Fourth Avenue.

Rick's telegrams were store orders for manufacturers and wholesalers. At a time when Alaska's long-distance telephone calls were obscenely expensive, Nerland's Home Furnishings communicated with its suppliers this way, sending the telegrams out at the close of business one day and expecting to get order confirmations the next. There were no fax machines, and creation of personal communication services and equipment like cell phones with email, texting, and Facebook access, lay decades in the future.

Rick never made it to the telegraph office. Fascinated by cameras, he stopped to window-shop in front of Stewart's Photo Store, a pioneer business still located in downtown Anchorage some fifty years later. Staring through the store's large plate-glass window, he admired state-of-the-art Rolleiflex and Hasselblad cameras.

Suddenly the window began shaking. Then the sidewalk under his feet began moving. Rick instantly recognized it was an earthquake, having seen the movie, *The Last Days of Pompeii.*

"I thought, 'Oh, my gosh, what if the ground opens up, what happens?'" he recalled.

Rick was about to find out. Glass in store windows along Fourth Avenue began trembling, then bursting. He moved to the edge of the street, between two parked cars. A stranger passing by grabbed him and walked him into the middle of the street, no easy task because the street was shaking so violently. They had to walk in a crouch to avoid falling down.

The memories lodged in Rick's mind for decades are as vivid as if he had taken photos or made movies of the scene. He was a fortunate witness to history in the sense that he was not killed, but Fourth Avenue, the heart of Anchorage's business district, then and now, began falling to pieces in front of him as he held his footing in front of Stewart's.

Rick had his eyes on the Anchorage Westward Hotel as a crack formed in the façade.

"I actually watched that crack start on the left, and what would be the west corner, and progress to the east corner," Rick said. "And then the shaking seemed to kind of settle down a little bit."

The respite was brief and then the shaking picked up more power, as if a deep breath was being exhaled. A block away Fourth Avenue was tearing open, swallowing cars, and sucking parts of buildings that suddenly dropped below sidewalk level.

The roar of the earthquake and the sounds of ripping concrete competed with the screams of a woman who drew Rick's attention. Wearing high heels, the woman seemed as graceful as a Broadway dancer, high-stepping her way down the sidewalk. Storefront glass exploded from their frames, landing in shards all around her.

"The glass panes were falling out and crashing at her feet," Rick said. "She made it about three or four car lengths and then darted into the street and was never hit by glass. It was amazing that she never got hit."

The city of Sitka in southeast Alaska, the former capital of Russian America, had donated a flagpole to Anchorage that was erected at the corner of Fourth Avenue and E Street. Rick watched with amazement as the flagpole "whipped around like a car antenna. It would probably go back and forth forty-five degrees and the concrete base kind of made it look like a bobbing cork."

For more than the five minutes of the most powerful earthquake in U.S. history, Rick balanced his feet in the middle of the street, shocked by the scene of constant destruction. When the quake stopped, he was surrounded by broken glass and saw cracks in many of the buildings around him.

Yet Rick was alive. He desperately wanted to get back to the family store but wasn't sure how to get there safely. The first thing he decided—and this was an eleven-year-old boy thinking—was as sensible as an adult choice. The telegrams could wait. He began negotiating a path back to the store through the turbulent mess. Downtown had been hit hard.

"There were cracks in the pavement and a little bit of mud and water was gurgling up from the broken mains, but interestingly enough I walked down to the corner of Fourth and E, and I never saw all of that slippage of the road a block way," Rick said. "But what I did see was a mailbox that had been bolted to the cement lying on its side. I looked at that and that's when it occurred to me how really powerful the earthquake had been, because it was huge. You can imagine a bolted mailbox ripping out of the concrete. That's a lot of power."

At the Nerlands' store, its floor lined with debris from broken goods, the family and workers stepped outside to the sidewalk. What faced them was the collapsed wall of the new JC Penney building, a tremendous hunk of concrete and metal that had separated away from the structure and fallen on parked cars, crushing them. In some cases, people were killed. In other cases, drivers were trapped but later rescued.

Alaska is known as the Land of the Midnight Sun and for minimal natural light in the winter. The longest day of the year, December 21, is dark twenty-four hours in some Arctic communities. In Anchorage on that day there is about five hours of daylight. On June 21, the longest day, the sun stays above the horizon for less than five hours. Clock-watching is a common component of Alaskan life, and the amount of sun drives certain elements of daily existence. In late March, 1964, a week into spring, there were about twelve hours of daylight.

Inside the furniture store, however, with the power out—as it was all over town—the Nerlands couldn't see much at all. Steve wondered if there were any emergency lights, but the family wasn't hanging around the store after the big shake-up. Roger was retrieved from under a dining room table on display in the basement.

Jerry Nerland and others nailed sheets of plywood to cover broken windows, and then Jerry set out toward the post office to find his son Rick. They bumped into one another on E Street, where Rick had gotten his first glimpse of the JC Penney store that was to become a symbol of Anchorage's downtown destruction.

"Up until that point I hadn't seen that kind of devastation," Rick said. "I knew glass had fallen out. I hadn't seen the Fourth Avenue slide-down, but walking around that corner, that was an 'oh boy' moment."

By the time Rick and Jerry returned to the store, the rest of the family was outside waiting for them, anxious to go home. Everyone piled into the family station wagon and began driving to their rented home in the Inlet View area not far away.

The normally short ride was informative, illuminating, and depressing. The route took the Nerlands past the new Four Seasons apartment building that had been under construction and was now leveled. They were starting to gain an appreciation for the magnitude of the damage. Jerry Nerland wondered aloud if the Emergency Broadcast System had been activated and if they might pick up news on the radio. But it was too soon—no such information was coming through yet.

The drive was laborious and cautious. Parts of streets that had been straight had become "serpentine," as Rick put it. But they were passable.

"Everybody was moving pretty slow," he said. "There were some pretty good cracks in the street where the street had opened up and were hard to drive around."

They reached the house. Their home was still standing. The Nerlands reclaimed four-year-old Brian from a very nervous babysitter. She was holding Brian tight, still crouched in a doorway. Both seemed to be in a state of shock.

Despite their relief that the house had not experienced major damage, broken dishes mostly, they were shocked by the view out the window. The Nerlands could see to Point Woronzof, where trees were bent at odd angles and homes had tumbled down hillsides.

A neighbor said he heard on the radio a warning for residents to turn off the gas. Gas lines might have ruptured, and there was fear of an explosion.

The entire Nerland clan gathered in the cold living room for the night. Sleeping bags were dug out and placed in a circle around the room as Jerry lit a blaze in the fireplace. There was a worry about aftershocks, and they did hit, jangling already frazzled nerves and rekindling fears. Anchorage residents were frightened, concerned that an aftershock might cause more damage and casualties. Everyone was on edge that night, especially with so little information. Facts were hard to come by. Nobody knew how widespread the damage was. Nobody knew how many were dead.

During the night a powerful aftershock woke Steve and sent him sprinting to the door. In his groggy state, he remembered being told there was relative safety standing in a doorway.

Jerry got a battery-powered radio going and listened to news updates. Official updates were vague. There was talk about tsunamis. Word was somehow spread to the Lower 48 that Anchorage had been wiped off the map. Was it true? It didn't seem to be, based on what they had seen in their drive out of downtown. It was difficult to assimilate news that you hadn't seen with your own eyes.

"It was frightening," Jerry said years later.

Radio broadcaster Jeannie Chance spent sixteen straight hours on the air trying to serve as a focal point for information. The Nerlands did not make it through the night at home because it was announced that a tsunami was coming and would endanger people and homes in their neighborhood. Members of the Anchorage Fire Department went door to door reinforcing that message. The Nerlands packed up and went to the home of Ted and Ann Stevens. Stevens was three years away from beginning his long career in the U.S. Senate.

By then word had spread about the quake devastation and a tsunami hitting Valdez and Seward, so the warning seemed quite real. The Stevens lived about four blocks away, and that's where the Nerland boys awoke the morning after the quake.

Anchorage was without power. Temporary water stations were being set up. The military was being called in to assist rescue efforts and prevent looting. Downtown was closed off. Most people, like the Nerlands, were trying to piece together exactly what happened, who was hurt, who might have died.

Alaska doesn't get tornadoes. In some areas of the state it never gets humid enough for thunder and lighting. The state doesn't get hurricanes. The one natural disaster that Alaska is ripe for is the big earthquake. That nightmare had become reality.

2

Turnagain

Turnagain by the Sea is a well-to-do neighborhood with fancy homes built close to Cook Inlet by many of Anchorage's prominent families. *Anchorage Times* publisher Robert Atwood lived there. Pilot Lowell Thomas Jr., son of the famed adventurer and writer, and later lieutenant governor of Alaska himself, lived there with wife Tay Thomas and their children.

When the earthquake shook the city, chunks of streets in the heart of the downtown district were churned into rubble. Buildings in the shopping area of downtown fell to pieces, or shed huge cement blocks from their walls or roofs. Some were flattened. But the destruction was worse in Turnagain, where homes were flattened, thrust off their foundations, shaken into slabs of concrete and jagged pieces of wood. Entire rows of homes tumbled down a cliff to water's edge, and then, when tsunami warnings were issued, it was feared the remnants and the neighborhood might be washed away.

Lowell Thomas spent most of his adult life as a bush pilot, ferrying mountain climbers to the glaciers adjacent to Mount McKinley so they could attempt to scale the 20,320-foot mountain that is the tallest in North America. A distinguished, deep-voiced man whose words carry the timbre of a broadcaster,

Thomas conducted business around the state by flying himself to distant communities in single- or twin-engine planes.

On the afternoon of the earthquake, Thomas was flying in his Cessna 180 to Fairbanks, 360 miles north of Anchorage, where he was scheduled to give a political talk to a group of Republicans that evening. After taking off at about 3 p.m., Thomas did not hear anything about the disaster until he landed in Fairbanks.

As soon as he heard the news, Thomas wanted to turn around, but commercial airports in Anchorage were shut down. The air-traffic control tower at Anchorage International Airport had collapsed. Thomas parked his plane and talked his way aboard a relief aircraft stocked with supplies being flown by famed pilot Merrill Wien to Elmendorf Air Force Base near Anchorage. Thomas was oblivious to the fate of his wife Tay, son David, six, and daughter Anne, eight, and his home.

Landing at Elmendorf about 3 a.m., Thomas noticed cracks on the tarmac. The lights were out.

"The big question was whether the runway was okay," Thomas recalled many years later.

Thomas learned that his family was alive, but the family's home was not.

"The house was a total loss," Thomas said.

Tay Thomas, reporting for *National Geographic*, wrote one of the most detailed accounts of the Good Friday Earthquake and its immediate aftermath. Her story appeared in the July, 1964, issue under the headline "An Alaskan Family's Night of Terror." At a time when American society was more conservative in its outlook toward working wives, Tay's byline read "Mrs. Lowell Thomas Jr."

In her account, Tay recalled that the weather was still wintry and that snow had fallen over the previous two days. Still, the family was not dressed very heavily indoors as it gathered to watch television. David was barefoot. Tay took note of the time when she felt the first rumbling of the earth just after 5:30 p.m. The entire house began to shake soon after, and the trio fled out the front door into the street. The Thomases were about ten feet beyond the house when, Tay reported, "it suddenly seemed that the world was coming to an end. We were flung violently to the ground, which was shaking up and down with the sharpest jolting I've ever felt. It seemed an eternity we lay there in the snow."

Then they witnessed their house being shaken apart, the windows shattering, boards being splintered, the garage roof falling in on their car. Around them the earth split open. The house and the land, with them on it, were heaved down an incline to the edge of Cook Inlet from its perch on the top of a hill. The children were crying hysterically, but mom calmed them, urging them to pray to God for help and survival.

The three Thomases managed to make their way back up the cliff, taking between fifteen and twenty minutes for the short, but challenging, climb. A man saw them from above and gathered up a group of people to help lift them to safety. Meanwhile, the home of Dr. Perry Mead, the Thomases' next-door neighbor, was wrecked. Five children lived in the home, where two of the young boys perished, twelve-year-old Perry and a baby brother.

The Thomases were wrapped in blankets and hustled to a nearby home for care. It was feared David might have suffered frostbitten feet, but they were warmed up adequately. Later that night, as radio communication improved, Tay Thomas learned of the widespread damage in the outlying communities of Valdez, Seward, and Kodiak, and in downtown Anchorage only a few miles away.

Once Lowell Thomas was reunited with his family, he and Tay tried to salvage their belongings. They crept out over the debris to their ruined home and retrieved suitcases of clothes and prized possessions saved from years of traveling. Their furniture was crushed.

The Thomas family rented a home until it could move into a new home a year and a half later.

AL BRAMSTEDT JR., who later became a prominent Anchorage radio and television broadcaster, grew up in Turnagain. When the quake hit, Bramstedt, thirteen, was playing outside with his close friend Jimmy Woodard at the other boy's house.

The Woodard home was located on the ocean side of Iliamna, at a T in the road where Bramstedt and Woodard were building a snow fort. The snow was fresh from the day before, but wet because the temperature was not intensely cold. The combination allowed for good building techniques. The boys could lock in the chunks and then pour water over the exterior so it would freeze solid in the declining temperatures overnight. That made for a strong structure the next day.

They knew precisely the routine to follow because they had built many snow forts. Woodward was watering the snow fort when the ground shook and tipped him over, knocking the water bottle out of his hand.

Growing up in Alaska, the boys were used to earthquakes, so their first reaction was to laugh when Jimmy was thrown to the ground. The summer before, Bramstedt recalled, an earthquake that lasted for about thirty seconds had been strong enough to spill a can of paint that was perched on a ladder.

"We had been through a lot of earthquakes—no big deal," Bramstedt said. "But this one was different. It was super-strong and it wasn't little vibrations of movement, two- or three-inch vibrations. This was more like east-west. It would go one foot one way and one or two feet the other way and then one or two feet back the other way; back and forth and a lot of motion."

The boys figured the quake would be over in seconds, but it went on and on.

"Then you've got adrenalin going and every minute seems like ten minutes," Bramstedt said.

Something to remember was the noise. The earthquake made its own sounds and created other noises. The spruce trees by Woodard's house were tall and there was no wind, but they began whipping back and forth as if being buffeted by one-hundred-mile-per-hour winds. The sway, he figured, was six feet to either side. They were truly rocking and rolling.

"They were making their own wind!" Bramstedt said. "That was something I'd never seen before."

After a minute or so of shaking, a new sound was heard coming from Cook Inlet. What they were hearing from a few blocks away were houses tearing away from their foundations, coming apart at their seams, and crashing down the hill towards the water.

"The backyard of the next street over was where the slide stopped," Bramstedt said. "It got louder and louder. Imagine the sound of a big, laminated beam, one and a half to two feet wide, just being broken, being crunched. But we never imagined what was really going on. We weren't seeing any damage at all. We could hear it, though, and I turned and remember thinking, 'What is making all that noise?' It was an incredible sound, glass exploding, wood splintering, groaning and crunching."

It was the slide that took down the homes in Turnagain by the Sea.

"It was an eerie feeling," Bramstedt said.

That creepy feeling was more pervasive in the high-strung or in those who might not have had Alaskan earthquake experience. Some might have thought that the end of the world was at hand.

Bramstedt moved into the street, where he saw a woman who had been house-sitting for a neighbor come running down the street. As she ran toward the two boys, she shouted, "It's a nuclear war! The Russians have dropped a bomb! We're all going to die!"

"She was absolutely panic stricken," Bramstedt said. "I remember looking over to the sun that was in the west and I thought, 'Well, that's the sun. That's not a nuclear bomb. This is just a whacked-out adult.' We didn't pay any attention to her."

Abruptly, as a crowd formed, the street split open about two feet away from Bramstedt, the blacktop asphalt suddenly resembling a dark crevasse of the type one would see on a glaciated mountain. It opened fifteen feet wide and came close to swallowing Bramstedt.

That is something Bramstedt has always remembered in detail, but at the time he said he was strangely calm, not scared, realizing that the earth was coming apart, but never thinking he was going to die. A young teen then with no responsibilities, he recalls it seemed like a grand adventure.

"It was an interesting emotion," Bramstedt said years later. "It was like, this is history happening and the most incredible earthquake imagined."

Meanwhile, Al Bramstedt Sr. was in Tokyo that day on an Alaska trade mission along with future Governor Walter Hickel. Young Bramstedt's mother, Rosa Lee, and two of his three sisters, were shopping in downtown Anchorage, where they witnessed Fourth Avenue rolling up and down before breaking apart and the family's 1963 Mercury bouncing into the air.

Meanwhile, the third sister, sixteen-year-old Susan, was with a friend in a Volkswagen bus that was heaving up and down. She managed to pick up Bramstedt from his friend's house, and he recalls that driving the mile and half home was like running an obstacle course. Every forty feet or so, they encountered a blacktop crevasse that they had to get around.

A huge stone chimney was a defining characteristic of the Bramstedt home. It had collapsed onto the roof of the garage, which had been converted into a family room.

On Fridays, a woman came in to clean. When Al and Susan arrived, she was sitting on a couch drinking their father's Cutty Sark out of a coffee mug. The woman just sat there sipping the Scotch. She was in shock.

The miracle was that the heavy stone chimney hadn't smashed the garage roof. The structure held up better than anyone could have imagined. There was some foundation damage, and the kitchen was a mess. All of the oils and liquids, powders, and sauces were thrust from cabinets and onto the floor, mixing with glass from broken dishes and glasses.

Bramstedt was assigned the task of cleaning it up with a shovel. He donned rubber boots as protection against the glass, gliding along like an ice skater over the floor that was slick with cooking oil.

"I would just slide along to the big trash cans," he said.

More surprising was the endurance of an heirloom in the living room. When they converted their home to electricity, Bramstedt's grandparents Oscar and Ruth gave his parents an eighteen-inch-tall kerosene lamp on a heavy brass base. Apparently it had slid back and forth on a shelf but didn't break.

After some debate, it was decided that the foundation of the house was skewed too much for the family to stay in it safely that night, so the Bramstedts evacuated to the home of a friend, Morgan Richardson. However, soon after they transferred to that home, a rumor circulated that a colossal tidal wave was headed for Anchorage.

Richardson protested, saying it was not possible. Don't worry about that, Bramstedt remembers him saying. At the time, the boy was looking out the window, watching people flee the neighborhood in cars and on foot. His mother said, "Are you sure?" And Richardson said, "It's not going to happen." As the panicked neighbors disappeared from sight, Bramstedt thought, "I sure hope he's right."

Richardson's analysis was correct. A day later, Bramstedt prowled the neighborhood, hearing firsthand about the slides, the damage, and the deaths of the Mead boys. Perry was a classmate whom he liked very much. When the quake hit, the family made a run for it, leaving the baby behind in the house. The twelve-year-old ran back in to get his brother and was working his way back to the family when the earth opened with one of those fearsome crevasses, and they fell in. Their bodies were never found.

IN TOKYO, Al Bramstedt Sr. had been on a telephone call to Anchorage when the Alaska party exclaimed, "Boy, we're having a really bad earthquake. I'll call you back." The return call never came.

The way the younger Bramstedt told the story passed down from his father was that Walter Hickel, who died in 2010 at age ninety, kept dialing but could not make the international connection. The phone lines were out all over Anchorage. The Alaska businessmen began picking up fragmentary reports indicating Anchorage had been consumed by fire. They were desperate to make contact.

Out of exasperation, Hickel supposedly impersonated a general on the telephone, telling a long-distance operator it was imperative that he get through to Anchorage. He did and received a more accurate picture of the damages, one that was quite serious, but not Armageddon, either.

So much was going on in so many places, from downtown Anchorage to Turnagain, Valdez, Seward, Kodiak, Whittier, Seldovia, and elsewhere that it was impossible to obtain accurate information. Initial reports put the magnitude of the earthquake at 8.6. Years later, after much study, seismologists upgraded the reading to 9.2, the second-most powerful in history. But with Alaska being an isolated state and most communications down, the only way to find out about what was happening somewhere was to see it with your own eyes.

ROBERT ATWOOD was born in Chicago in 1907 and moved to Alaska in 1935, when Anchorage had about 2,500 people. Aided by his wealthy father-in-law banker, he was able to buy the *Anchorage Times*. As editor and publisher, he built it into the largest newspaper in Alaska. Although the *Times* eventually lost a newspaper war to the *Anchorage Daily News* and went out of business in 1992, Atwood long used the paper's editorial pages as his bully pulpit. There was no more devoted and expressive booster of Alaska, especially Anchorage, than Atwood.

He campaigned enthusiastically and vigorously for big-dream projects such as getting military bases located in the territory, drilling for oil, and winning statehood for Alaska. When the U.S. Senate approved Alaska as the forty-ninth state on June 30, 1958, a gigantic, six-inch-high headline on the front page of Atwood's newspaper proclaimed:

WE'RE IN

Atwood's life, profession, and the earthquake were on a triangular collision course. The earthquake erupted with such force and surprise that his newspaper was unable to publish an edition on Saturday, March 28, 1964, though it did put out an "Extra" on March 29. Meanwhile, Atwood's home in Turnagain was destroyed.

Atwood, who died of heart failure in 1997 at age eighty-nine, was three days shy of his fifty-seventh birthday on Good Friday. He was alone in his well-appointed house in the hardest-hit area and had to run for his life. One of those yawning earth crevasses nearly got him. He had to fight his way free. Atwood was practicing his trumpet when the earth moved. It was the only thing he took with him when he fled.

In the *Times'* extra edition, Atwood wrote of watching his house disintegrate and then nearly losing his life.

"On the driveway, I turned and watched my house squirm and groan as though in last mortal agony," he wrote. "A chasm opened beneath me. I tumbled down. I found I couldn't pull my right arm from the sand. It was buried to the shoulder."

Atwood let the trumpet go and saved himself.

MARY BURKHART and her late husband, Terry, were living in the Turnagain area, although not in one of the spiffy, expensive homes. From home they operated an apartment-refurbishing business called Burkhart Furniture and Upholstery but were in the process of moving. They had sold their house and were about to move a little farther inland into a friend's fourplex. The Burkharts had been out of town on an extended visit to Hawaii for golfing and a visit with Mary's sister.

They chose to fly home on March 27, planning to finish their move that day. Mary, thirty-six at the time, recalls their flight landed about 2 a.m. Later in the day Mary bought a pot roast and a bottle of red wine for dinner at their new home.

The pot roast was cooking, and the Burkharts and another couple were sipping wine when the shaking began. Recognizing immediately that it was an earthquake, Mary, who said she was naturally bossy because she was the oldest sibling in her family, started giving orders.

"I started screaming, 'Get in the doors!' Well, there was only one door to their apartment. So we're all trying to get into that one doorway."

After a few moments of that overcrowding, everyone decided it was best to get outdoors. Mary had read a book about the demise of the world and thought it was coming true. "Oh, my God," she thought, "the world is coming to an end."

The stove tipped over, and the pot roast slid across the floor as if it was a fumble in a football game. China began shaking free of the cabinets, one plate at a time, crashing on the floor. The refrigerator stayed upright but was wiggling away from the wall.

Outside things were hardly more reassuring. The violent swaying of treetops back and forth astonished the couples. The land was rippling in two-foot waves.

"The ground was swelling like that," Mary said.

Some of the swaying trees did not straighten up all of the way again, and some broke. Mary decided getting into her car and driving away might be a safe way to go. She waded through some snow, barely able to stand up and tried to climb into the car, a green Thunderbird with a white top. Her dog, a little brown dachshund named Kurt, was jumping up and down next to the car door.

The Burkharts pulled the door open, the dog leaped inside, and she started the car. The other couple got into their car, too, and someone yelled, "Head to the mountains!" That meant the Chugach Mountain Range, whose peaks just beyond the city's eastern outskirts ranged in height from 3,500 to 5,300 feet—well above the reach of any tidal wave.

Getting out of town was a challenge. Motorists did not go far before realizing it was not going to be a straight drive to the hills. They soon were confronted by large, gaping holes in the roadway.

"We would have to drive around big cracks in the highway that were very, very scary," Mary Burkhart recalled many years later. "Trying to get out of town to the mountains, I don't know how many streets we tried ... When we picked up information on the radios later, they said we would have a tsunami in Anchorage and we should stay at higher ground."

Hours passed before the Burkharts heard on the radio that no tsunami was going to clobber Anchorage. They returned to the fourplex in the middle of the night. The damage was minimal. The power was out. And the pot roast was tossed out.

AS IN MANY natural catastrophes, some lose their homes while others are untouched. Betty and Gene Johannes owned a place on Illiamna Avenue in Turnagain, close to where Al Bramstedt Jr. and his pal built their snow fort.

The quake knocked down a hill some houses situated on the cliff close to the water and rearranged the neighborhood landscape. The Johanneses' home was surrounded by fissures and cracks in the front and back yards but came through virtually intact.

"We ended up with a bluff lot," Gene Johannes told a newspaper reporter in 2007.

3

Downtown Anchorage

Mike Janecek was a fifteen-year-old sophomore at West High School who had the day off from school. Like the Nerland brothers, he was putting in a day at his after-school job.

Janecek worked in the Fourth Avenue Theater building for KENI-TV Channel 2 (later to become KTUU-TV) and KENI radio. Because he was too young to drive, Janecek was picked up by his parents after work, but his job carried considerable responsibility.

Later in life Janecek was a high school teacher and athletic director at Palmer High School, about fifty miles north of Anchorage, but in 1964 he was a cameraman for a television station. His job was to film events for the 6 p.m. news broadcast.

At 5:36 p.m., Janecek stumbled into the chaos in downtown Anchorage. Janecek had been ill with the flu the day before. But he felt a bit better by Friday afternoon and showed up for work.

Believing there was nothing a good chocolate milkshake couldn't cure, Janecek was on the stairs from the basement radio station to street level, heading for the Woolworth's fountain. He didn't recognize the rumbling noise at first, and then the building trembled.

"I thought, 'Oh, gosh, I'm going to be sick again,'" he said.

The shaking became more violent. The stairs and walls were moving.

"Well, within seconds as I got up there, the shaking was more obvious and I realized, 'All right, this is not equilibrium, this is some problem.'"

When he emerged onto Fourth Avenue, the wall of Adams Stationers nearby was crumbling.

"It had me spellbound," Janecek said. "I was staring at this when the window of the building exploded, the glass came flying out. There was dust coming out of the window and out of the hole comes a kid who was working there."

Janecek, a lifelong track and field fan, instantly thought of a running analogy as he watched the young man zoom through the empty space that had held the window, which had shattered. "That kid has perfect hurdle form," he thought. The kid kept on running down the street while Janecek, who had been rooted to the spot, decided the kid had the right idea.

"That's it!" I thought. "Run!"

Janecek ran across the street to the post office, which was surrounded by a chain-link fence, and grabbed onto the fencing. Others did the same thing. The post office was decorated, if it could be called that, by ugly stucco walls. Janecek was growing up at the height of the cold war, during which he and fellow teenagers were bombarded by the contemporary anti-Russian information. They were raised on duck-and-cover dives under their desks as protection from a nuclear bomb blast. He and his friends made fun of the post office, calling it the Kremlin.

"It was this big, massive structure, but it was spitting off these little pieces of stucco," he said. "They were landing around me. It startled me, scared me, and a woman was wrapped around a nearby parking meter screaming at the top of her lungs for us to repent, that this is it, folks. Boy, it scared me. It really did."

A woman screaming about the end of the world unnerved Janecek. A big man wearing a black watch cap, dressed as if he had just come ashore from a fishing boat, was listening to the same hysteria. He looked down at Janecek and said, 'Don't worry, kid, it's just an earthquake.'"

"Just an earthquake" was not much reassurance given the amount of destruction taking place around him. After about three and a half minutes of ground vibrations, there was a lull. Janecek thought it was all over. He started to make his way back to the station.

"I only got about ten steps away from the fence and it started again, so I turned around and ran back to where I was, grabbed onto the fence, and I looked up at the guy. He looked at me and said, 'I don't know, kid. This could be it.'"

As the ground shook ever more powerfully, Janecek wasn't so sure his four-foot-high chain-link fence was much of a safe haven, after all. Cars that had been on the move screeched to a halt on the street, and the teenager became terrified when the street began splitting apart.

"You could see the waves of the earthquake going by and you could hear the asphalt crunching," Janecek said. "When it stopped, I waited for a minute or so because I didn't want to make the mistake to go again and then have to turn around and come back. I waited, then I took off and when I got diagonally across the street, the people from inside the theater came out."

The Fourth Avenue Theater had been filled with kids for a movie showing of *The Sword and the Stone.*

"They were crying and all sweaty, their hair messed up," Janecek said. "I found out years later that the guy who was the young manager jumped up on the stage with a flashlight, shined it on himself so the kids could see him, and said, 'Now, nobody move. If you all go through the door, you're going to get hurt. Everybody just stay right here. This is a really solid structure.' And everyone stayed right there until he released them."

The young manager was right. The theater did sustain much damage, but nearly a half-century later, it still stands.

Janecek knew he was going to have to go to work and lamented that he did not have his movie camera as the kids flocked out of the theater. Surveying the damage around him, he saw that the new tower under construction at the Westward Hotel was missing a piece. He could make a lot of money, he thought, if he could market the news film he wasn't taking.

Once back inside the station, Janecek was asked to accompany co-worker B. G. Randlet around downtown—not with a camera, but to make an assessment of what damages they saw, and to report back. Station higher-ups wanted a reconnaissance while they worked on damaged equipment and tried to get the stations up and running.

"We had a generator out there, and they were firing up the generator to get on the radio," Janecek said. "We found a steel beam lying in the street at Third and F. The bars had sunk down to below street level, but in some ways it looked the same as it did every day. People walking around in front of the bars with drinks, laughing and giggling, yet they were twenty feet down there. We got to the Empress Theater, or maybe it was the Denali Theater, and the marquee was sitting on the sidewalk."

On D Street they saw the JC Penney store. It was then that Janecek began to feel his mentality shift from viewing the quake as a bit of an adventure to a life-and-death worry. He knew people must have been killed at Penney's. Then he started to worry about his parents and his three sisters.

"I just wanted to get home," he said.

Janecek knew his parents weren't coming to pick him up downtown any time soon. The phones were out, and the only way to get home was to walk. He lived in the Spenard area, perhaps six miles from the Fourth Avenue Theater. When he and Randlet reported back about what they saw, the station manager assigned Randlet to take Janecek home. Randlet, twenty, a newlywed, got permission to drive past his own home. Except for a tumbled-over chimney lying in the yard, nothing seemed amiss.

Getting to Janecek's house was more complicated. The route normally would be up Romig Hill, but part of the hill had slid down, causing the road to buckle about twenty feet from Chester Creek. A four-foot pyramid of rubble and contorted pavement made the road impassable. Cars were backed up slowly trying to drive through the swampy area of Chester Creek, not very deep that time of year. The situation strained the patience of those anxious to reunite with loved ones.

When Janecek got home, he was greeted warmly by his mother, Catherine. Just as the quake hit, his father, Frank, who worked for the state highway department, had been dropped off nearby by a co-worker. Frank Janecek was hurrying the final stretch to the house when the ground movement caused him to fall, and he cut his face.

At home, as things were flying from shelves, Catherine rushed to protect her prized possession—a china cabinet containing all of her favorite trinkets.

She leaned into it, holding it against the wall, and held the cabinet doors from opening. However, at the same time, across the kitchen, the refrigerator began moving, with the door opening and shutting like the jaws of a beast headed toward her.

Mayonnaise and ketchup bottles broke, and sauces were oozing out of the fridge as it moved. Just then Janecek's father arrived on the scene, and Catherine shouted, "Frank, the refrigerator is alive!" He said, "What are you talking about?"

"That walking refrigerator," Mike Janecek said, marveling at the memory.

Stories of refrigerators on the move were countless during the quake. They were big and strong and lodged in corners against walls, but the shaking was so strong that they were relocated into the middle of rooms.

Young Janecek stayed home for a day, but when telephone service was restored, he called Al Bramstedt Sr., his boss, who asked him to report to work. He made his way over to Bramstedt's house, where he was picked up by news director Ty Clark. The radio transmitter for KENI was working because it was still standing on Hill Crest Drive, near West High, away from some of the most severe ground movement. A generator providing back-up power was housed in a building between the school and nearby salt flats.

Quickly, survivors began to rely on the radio station for news of what was happening and to try to get word about missing persons, or to let others know they were okay.

The city was stunned, in shock, and reeling. Much of downtown was devastated, and residents there were fearful. At first no one knew what hit them. There were a lot of conversations with God. The next thoughts were of self-preservation.

LINDA MYERS-STEELE, who at the time was nineteen-year-old Linda Henderson, was born in New York, lived in Palmdale, California, and moved to Alaska following her high school graduation to join her married sister and family. The arrangement was to stay for six weeks and find out if she liked Alaska. If she didn't, she was promised a free airplane ride back to California. It didn't take her long to decide. Soon she wrote her boyfriend in the Lower 48 a "Dear John" letter and has been in Alaska ever since.

Within ten days of her arrival, Linda got a $200-a-month job at Fred Kohli Motion Picture Services. She was part secretary and part booking agent for sixteen-millimeter films that were shipped to rural bush villages for entertainment along with bulk quantities of popcorn.

"Raw popcorn was ordered from the Carolinas, popped and packed in two-pound bags," she said. "Two pounds is a lot of popcorn!"

At the time of the quake, Linda Henderson was washing her hair in the bathroom sink of a basement apartment in Spenard that she shared with a roommate. Linda was bending her head into the sink when the first ground wave struck. As she lifted her head, the contents of the medicine cabinet spurted out and landed in the sink. It got worse from there.

"The shaking was fearsome and wouldn't stop," Linda said. "I could not walk steadily. I went outside and held onto a post. The tops of the tall birch trees literally touched the ground on one side, rose and touched the ground on the other side. It was as if they were feathers!"

Throughout the rumbling ground shifts, she was in shock, not realizing how big the earthquake was. A chimney had fallen in on the second floor. Inside, she found that everything stored in her cabinets had exploded onto the floor.

"That's when I realized that this was not a minor occurrence," Linda said.

Inside, Linda joined her landlord, who lived upstairs with her kids. The woman's husband was out of town. They turned on a transistor radio for news. Linda's roommate, Pat, arrived from Fort Richardson, where she worked, having weaved in and out of torn-up roads on her way across town.

"There was no electricity, no heat, no flashlights, just a few candles," Linda said. "We wrapped up in blankets as the night grew colder and stayed awake most of the night talking and reassuring the kids that we'd be okay. The aftershocks were very frightening," Linda said. It was a long night.

The man who rented a basement apartment next to Linda's returned home with a generous offer. His friend was a ham radio operator, who would deliver messages to the Lower 48. They knew exaggerated reports of Anchorage being consumed by fire would definitely alarm people. Linda was able to get a message to her parents in California that she and her sister's family were okay.

"I've had a special spot in my heart for ham radio operators ever since," she said.

Linda was never tempted to leave Alaska, but the Good Friday Earthquake changed her outlook about earthquakes.

"I respect the power of nature," she said, admitting that she felt lucky to have survived. "Luck, fate, God's plan for me—there was much to do in my life. It was apparently not my time to go."

AS A SECOND-GRADE teacher at Northwood Elementary School, Sarah Burkholder, twenty-five, had the day off, but she and her roommate had volunteered to go to the Elks Club on Third Avenue to help color eggs for a Sunday Easter egg hunt. They lived at Grandview Apartments a few miles away.

Burkholder and her roommate showed up about 4 p.m. at the Elks, where they joined other volunteers gathered around a bar. A mix of ten or fifteen volunteers and Friday-afternoon regulars were drinking.

They were waiting for the Easter egg hunt organizers to show, when the building started shaking. At first, everyone was calm. Earthquakes were as ubiquitous as snow in Alaska. Burkholder, who had moved from Buffalo, New York, in 1961, had toughed out quakes before.

"This one didn't want to give up," she said.

As the shaking got stronger and kept going, there was a let's-get-out-of-here moment when everyone began stampeding toward a stairway descending to street level. Burkholder noted that chivalry was nowhere to be seen.

"Some people had the forethought to take their drink with them, which I did not," she said.

Out on the street, parking meters and light poles were shaking, even as people hung onto them, and the sound of glass breaking could be heard from nearby stores and buildings.

Burkholder and her roommate clambered into the car, heading for home to the east, away from downtown and Turnagain, where most of the damage had occurred. On the way they encountered little road damage and at the time did not realize how serious the quake had been. At home it was almost as if nothing had happened. One cabinet door had opened, and a jar of mayonnaise had escaped. They cleaned up the glass, and that was it. But they had no electricity.

Their Friday evening had been rudely interrupted by a major catastrophe, which they were not fully aware of, so Burkholder and her pal decided to head for the Peanut Farm, then and some fifty years later, a popular watering hole on the south side of town.

The bar was open, but bottles had fallen from the back bar. Broken glass, liquor, and money spilled from a cash register mixed into a yucky and potent stew on the floor. A trap door to the basement, where most liquor was stored, was open and the operators were checking for damage. It was considerable. Perhaps a half-dozen people were at the bar drinking. Burkholder and her friend joined them, had one drink, and then were asked to leave as the bar was shut down.

The women returned home, where the landlord had made a fire in his fireplace and invited them to stay. However, they retreated to their own apartment and warmed up the radio instead. Jeannie Chance was on the air for KENI, serving up reports as she received them, including details about what happened in Seward, Kodiak, and Valdez. Gradually, it sank in that this was a disaster of the first order.

"We were just amazed," Burkholder said. "Turnagain was demolished, basically. Big parts of it were gone. Parts of downtown that we hadn't observed coming home were not there. JC Penney's was not there. We had skirted that area heading back to our apartment."

An uncle of Burkholder's, a ham radio operator in Buffalo, was able to contact her through another radio operator, who passed the word back to upstate New York that she was okay.

That's how it worked in many cases for worried family and friends thousands of miles away in the Lower 48 with no capability of reaching loved ones. Were they dead or alive? Was the city that they had chosen to live in, so far away, and so often against the wishes of less adventurous members of their clans, wiped off the map?

Every experience was deeply personal. Every experience was different. Many were similar, but most people didn't know that until later. Where were you when the Big One hit?

ERLDON "SPEED" GRADIOT moved to Alaska from Maine in 1951 and was working as an electrician with his father, Gus, at a one-bedroom, flat-roof house. They were in the basement fixing a boiler valve. The men, owners of Northern Electric Company, were about thirty steps underground when the rocking and rolling began. Gradiot immediately recognized what it was and knew they were in a dangerous place. “Let’s get out of here!” he shouted.

“Oh, it was a big shaker,” Gradiot said. “We had been through a few earthquakes. This was big.”

Speed got his nickname because he was born two weeks ahead of schedule and re-earned it scrambling out of that basement.

“Everything started rattling,” Gradiot said. “Book cases turned over. The chandelier was rattling. We knew this was different. It wasn’t three seconds after it started. When we hit the doorway … it was shaking so badly I couldn’t stand up. I went up the stairs on my hands and knees. I was driving a 1963 Chevy station wagon and we knelt down behind the bumper of that car. I watched that bumper go back and forth sideways five or six inches at a time. I thought the world was coming to an end. It just kept shaking.”

Gradiot kept staring at that chrome bumper as it shimmied. He didn't see a single person in the street.

“We didn’t move off our knees until it stopped shaking,” Gradiot said. “I don’t think we moved. We were scared to death. My dad and I didn’t say a word. We just looked at each other.”

When the shaking stopped, they climbed into the car and drove home to their Spenard neighborhood, where his father lived and where the shop was located. On the way, the Gradiots looked for property damage but mostly saw only collapsed chimneys. Gambell Street had settled a few inches, and the roadway was uneven in places.

The Gradiots stayed at the shop for a while, listening to news on a battery-powered radio. At the request of authorities seeking volunteers to help with rescue efforts, Speed and his younger brother Kris drove downtown to the JC Penney store.

“My dad had taught me to weld and how to use a cutting torch,” he said, “so we went to Penney’s and met the emergency people there. Oh, my God. I saw them pull a slab off the front of a vehicle and there was a woman in the car.”

That woman was killed by a falling hunk of concrete from the side of the building.

"There were two or three people there," Gradiot said. "The sides of Penney's were full of cars and the concrete slabs from the building flattened those cars … There was nothing there higher than a wheel rim. It was horrible. We volunteered to do some cutting and torch work."

For starters, Gradiot was asked to cut off the tops of parking meters to give a front loader access to the scene. Kris helped him. Gradiot pulled on his protective mask and began cutting. All of a sudden, he was interrupted.

"Get out of there!" he heard.

When rescue people on the scene ordered him to stop, he and Kris pulled back. Two large slabs of concrete were hanging off the Penney's building, and it appeared they could crash down on the brothers. The torching work was halted for a couple of days while a crane moved in to pin back the slabs.

CLEVEY COOPER OF EKLUTNA was in Anchorage at the D&D Bar, a floor below street level, when the earthquake got his attention. The twenty-six-year-old unemployed heavy-equipment operator was watching a pool game. A friend needed crutches to get around and couldn't move very fast. "Get out!" Cooper yelled to him as the game room evacuated.

The sidewalk and street were cracking when they reached outdoors. Nearby, the Denali Theater, which was playing *Irma la Douce*, was crumbling to the ground. Cooper said his heavy leather jacket protected him from the glass when bulbs and glass shattered.

"It filled my coat up with glass, but I was lucky, I didn't get any cuts or anything out of that," he said. "I never saw anything to compare with this. I worried that the ground would go out from under my feet, but it stopped."

The street caved in and took all the cars with it except Cooper's, which was the only vehicle left at street level where he had parked it on C Street between Third and Fourth Avenues.

"We had to get some wide planks and lay them down across this big ditch where the ground had fallen in, and push the car across them—that's the way we saved my car," he said.

He guessed that some of the cars dropped fifteen to twenty feet below ground level. After salvaging his car, Cooper drove home to Eklutna, where he found his home undamaged. His wife, Julia, and their three small children had been home. The kids were outside playing. When the earthquake hit, they were walking towards the house but lay down on the ground in a group until it was over.

"My kids were hollering, 'Stop it from shaking,'" Cooper said.

More than five minutes later, the shaking finally stopped.

MORT HENRY was a twenty-two-year-old student at the University of Alaska in Fairbanks in March 1964. He and his young lady friend had snuck away for the holiday weekend, coming to Anchorage without letting her parents know. After checking in at the Travelers Inn, they were headed for the Hofbrau restaurant for dinner. They looked forward to the all-you-could-eat crab legs' special and a pitcher of beer for four dollars.

The couple was strolling downtown to their destination when the earthquake hit with a violent force; the ground trembled and then broke open at Fifth and Gambell.

"We were frozen," Henry said. "We just kind of held each other. It felt like an hour."

They were rooted to the spot, too shocked to be scared, as the street moved in waves like the ocean. Yet, when the shaking ceased, the roadway was intact, unlike Fourth Avenue a block away.

When the earth settled, they retreated from the devastation. They had no clue how bad things were but decided to contact her parents. The couple began walking, but her parents' house was miles away. A Good Samaritan in a van told them an emergency shelter was being set up at Wendler Junior High, which was a much closer spot. That's where they headed. It was not the same as the hotel getaway planned, but they were out of danger.

CIRCUMSTANCES COMBINED to make Peggy Bensen's south-side Anchorage trailer and lean-to a noisy one on Good Friday. Her husband, Al, was sidelined with a broken leg. Her newborn son, John Robert, only five months old, was in a crabby, crying mood. She had taken the day off from her job as an accounts-payable clerk with the Anchorage School District, but she was unable to relax.

She cleaned the entire house, did the laundry and, like her husband, who did everything he could think of to pacify the baby, endured nonstop crying.

"He was really ornery," Peggy said. "He wouldn't even take a bottle. Al just rubbed his gums, gave him a little drink, and fed him with a spoon."

Nina, their little German shepherd, was acting nervous. As Al Bensen lay on the sofa, the dog uncharacteristically climbed up on him. It was as if the dog was looking for comforting. Then the earthquake arrived with a loud rumble.

"It was like a train coming," Peggy said. "I said, 'What's that noise?'" And then everything started to shake. Out the front window Peggy and Al could see a little neighbor girl, about eight, who had been on her way to visit the couple. "We had a small tree in our front yard and it was shaking back and forth, and Al could hear her yelling at it to stand still. She was hanging onto it," Peggy said.

Peggy sat down in a chair holding the baby, and Al was lying down on the couch, his leg anchored by a cast. Son Brian, age eight, sat calmly through the shaking, almost amused by it. Then the trailer and lean-to split apart, and suddenly Al saw the sky. Then the pieces came together again. Surprisingly, after the shaking stopped, the split had closed, perfectly matched in place, and Bensen swears it never even leaked after that.

Peggy had been cooking moose stew for dinner in a three-quart, cast-iron pot on the top of the stove.

"It just walked off the stove," she said. "I had to clean that up. It tipped over when it hit the floor, but there was little damage otherwise. There were a couple of broken dishes."

At first, Bensen thought it was a typical mild earthquake, but as it lasted longer and longer, she realized this one was in a category of its own. Despite the violent shaking, however, she said she was not particularly scared. Afterward, the Bensens were left without power, so Peggy retrieved candles and a gas lantern. The Bensens also had a shortwave radio a friend had left with them. It became a source of news about the quake and its aftermath.

Even baby John seemed to sense the gravity of the situation.

"He stopped crying," Peggy said. "I don't know if he could sense that something horrific was going on, or what, but then you would think he would have cried worse. But no, he stopped crying, and I don't think he cried again for two days."

JOHN SCHANDELMEIER, who long before abandoning Anchorage for rural living in Paxson, hundreds of miles north, was a sixth-grader at Abbott Elementary School on the south side of town. The twenty-seventh was his father, John Daniel's, birthday. In 1964, a house that far south of downtown was considered rural.

The Schandelmeiers lived with plenty of open space surrounding them. The house was on a homestead, abutting a construction worker's homestead. A log cabin more or less across the street was owned by another construction worker.

John recalls that the family had about 40 chickens, 25 ducks, 10 to15 geese, some turkeys, 2 beagles and a Labrador retriever, plus between 300 and 400 pigeons, living on the homestead. Despite speculation that animals sometimes sense the onset of an earthquake, Schandelmeier said the family's animals were completely silent, seemingly unaware that the ground was about to give way.

John, his father, mother Nell, and sisters Linda and Jeannette were just sitting down to dinner. His dad's birthday cake was on the table. At first the Schandelmeiers didn't react to the trembling. They were used to quakes. The reaction was "so what?" said John, who later became an accomplished dog musher and twice won the 1,000-mile Yukon Quest.

However, the shaking didn't subside. The power of it became evident when cupboard doors began bursting open. Everyone sprang to their feet and held the cabinet doors shut until the quake was over.

"It did feel like a long time, and when it stopped we all realized this was not just another normal earthquake," Schandelmeier said. "My dad said something like, 'Wow, that's got to have caused some damage downtown, so we'd better see what's going on.'"

The phone lines were dead and the power had gone out, as if the family needed more clues about the impact of the quake. But the birthday cake survived intact.

John the elder revved up the station wagon, a Chevy Bel Air, and he and his son worked their way past where the Dimond Center Mall stands today and through the Sand Lake Road area. Then they came across a large crack in the road.

"It seemed like it was ten feet wide," said Schandelmeier, recalling a twelve-year-old boy's amazement.

People had begun placing heavy wooden planks across the chasm. Cars worked their way across slowly and carefully, and the Schandelmeiers drove onto the Seward Highway and headed downtown.

A hotel being constructed on Seventh Avenue had collapsed, and a worker was killed. They saw the Penney's store but could not get close to Fourth Avenue. That night and on ensuing days, the elder Schandelmeier took pictures of the city's distress. They captured images of the community in agony. Eventually, his son came into possession of them after they had been stored in a box at the back of a closet for a very long time.

The family slept at home that night, all of the children in rooms in the basement. Before they went to bed, Nell warned the kids to run out of the house or stand in a doorway if an aftershock shook the house violently. Aftershocks did strike through the night, and Schandelmeier recalls being nervous. They were not so severe as to send the family into the street, but the Schandelmeiers spent an uneasy night.

4

Away from Downtown

The Anchorage of 1964 was a different, less-developed city than it is now. It was not uncommon then for a dog musher like Dick Mackey, who later helped organize the first Iditarod Trail Sled Dog Race, to live within the city limits with his kennel.

Mushers are an extinct species in Alaska's largest city now, but in the 1960s Mackey lived on the sparsely populated south side of town, where none of the neighbors complained about howling huskies. Mackey, who came to Alaska from New Hampshire, eventually gained fame for winning the 1978 Iditarod by one second in a sprint to the finish in Nome.

Mackey was working as chief operations timekeeper for the Alaska Railroad at the time of the Good Friday Earthquake. He had completed his day shift and arrived home for his nightly run with the dogs. Son Rick, age ten, who also became an Iditarod champion decades later, used to help line up the dogs for his father's run. Rick greeted his father at the door with a fistful of harnesses. The plan was to transport dogs to Eagle River, about fifteen miles away, and run a team from there.

"All we had to do was load the dogs," Dick Mackey said. "And that's when the earthquake struck."

Mackey felt the initial jolt in the kitchen, where the refrigerator slid across the room and bumped into the opposite wall. That was an eye-opener.

Stunned, the Mackeys looked outside and saw their trees lying perpendicular to the ground. Dick's first thought was that the Soviets had attacked Elmendorf Air Force Base, about five miles away. Transfixed, he watched the trees flip back, as if being catapulted, and swing 180 degrees until their tops touched the ground in the opposite direction.

"And then you realized it was an earthquake, of course, because the ground was shaking," he said.

Typically, when huskies see the mushers carrying harnesses, they erupt in excitement, eager to run. The dogs had reacted that way when they saw Rick gathering the harnesses. However, as the lengthy, violent earthquake played out, the dogs ceased barking and retreated into their small, wood-box homes. Afterward they didn't make a sound.

Dick, wife Jo, sons Rick and Bill, and daughter Becky were all home. Once the refrigerator did its dance and Dick realized that the nation was not at war, he shouted for everyone to abandon the house. The Mackeys owned an old Ford that had slipped down the driveway and partway down an embankment. The five of them lay down on the ground, holding onto one another as the ground shook hard. Rick said they ended up in a circle, arm to arm.

"The kids were crying, maybe not Rick, and the wife was crying," Dick Mackey said, "and we were all terrified. I made the comment to my wife, 'I don't know how much more this is going to take.'"

A fissure opened in the driveway, about three feet wide and forty feet long, "and then it slammed shut again," Mackey said.

Nearly a half-century later, Rick Mackey can still picture the way the quake moved the ground in waves, some three or four feet high. The car began moving again, and they worried it was going to bounce right over on top of them all. Those trees, their tops swinging back and forth, form an indelible image in Rick's mind.

Finally, the ground stopped moving, and everyone stood up and retreated inside the house to assess the damage.

"It looked like a herd of buffalo had run through there," Rick said.

The Mackeys had dug a ninety-foot well on their property. The earth shifted so dramatically that it disappeared. Dick Mackey said it was never found again.

The power was out. A family across the street with a fireplace invited the Mackeys and other neighbors into their home to keep warm.

"I remember being very scared," Rick Mackey said. "And then we had an aftershock. It was, 'Here it comes again.' Talk about one scared little kid. It was like a grizzly bear staring me in the face. It was pretty damned spooky."

The Mackeys turned on a battery-powered radio for news bulletins. One requested that people with first-aid experience go to Providence Hospital to help. Dick Mackey had been on a rescue squad in New Hampshire, so he made his way to the hospital.

It was chaotic," he said. "A lot of people injured, but they had been hit with stones and glass and just had bruises. It didn't amount to much."

ANOTHER OF THE FIRST families of dog mushing was much farther away on Good Friday. Joe Redington Sr., later to become known as the Father of the Iditarod, lived on a Flathead Lake homestead fifty miles from Anchorage.

"I think we were cooking dog food," said son Raymie. "The trees started swinging. It was pretty exciting. It seemed like it lasted quite a while."

The Redingtons paused from doing the chores and went into their house to try to make contact with Anchorage. The violence of the quake told them something big had happened. They were in remote territory, unlikely to see passersby any time soon.

Born in Oklahoma, Joe Redington had come north by way of Pennsylvania after World War II. He and his family had experienced many smaller earthquakes but this was different.

"You could really tell this was strong, really strong," Raymie Redington said. "The dogs were running around in the trees, scared. It was big. There were tremors afterwards, too."

The Redingtons made contact with Anchorage by two-way radio.

"They were saying there were torn-up buildings, that homes were sliding into the Inlet, and that it was pretty bad in Turnagain," Raymie said.

By then their dogs had calmed, and they realized the only visible damage around them was some fallen trees.

The epicenter of the earthquake was under Prince William Sound near Valdez, 305 road miles from Anchorage. By air it is much closer. The initial Richter scale reading of between 8.2 and 8.7 was later revised upward to 9.2, a tremendous, forceful impact that felt like explosions to many.

Southcentral Alaska took the brunt, but the shaking was felt in many directions. There were reports that some impact was felt as far away as Louisiana. Alaska is 586,000 square miles in size, and the principal cities are far apart.

FAIRBANKS IS THE second-largest city in Alaska. Signs on the Parks Highway say it is situated 360 miles north of Anchorage. Damage there was minor.

"It gave you a jolt, even up there," said Mike Doogan, who grew up in Fairbanks and later became a columnist for the *Anchorage Daily News,* a mystery book author, and a state legislator.

Then fifteen, Doogan recalls he was walking down the street at 5:36 p.m., having left church, when he felt the earth tremors in Fairbanks.

"You didn't have the instantaneous communication then that you have now," Doogan said. "So it took a while for people to figure out what it was that had happened. People got information in dribs and drabs. People didn't know how badly southcentral Alaska had been hit. There was a lot of mobilization."

The discussion in Fairbanks, as Doogan recalls it, was what role the city should play in terms of rescue work and being of assistance if people in and near Anchorage lost their homes and needed food or shelter.

"There was concern about what are we going to do if there were going to be a bunch of refugees, and where are we going to put them," Doogan said. "For a day or so, until things became clearer, people were making a lot of contingency plans."

Most of those plans were discarded quickly because although, as Doogan put it, Anchorage had suffered a large amount of "cosmetic" damage, most buildings withstood the earthquake, and circumstances there were not as bad as first reported.

TODAY, THE SAND LAKE area of Anchorage is thick with fine homes, but in the 1960s it was rural. The Sobolesky family was able to purchase ten acres. Karen Sobolesky, whose family had migrated from Washington State to Sitka and then to Anchorage in the 1940s, was twelve in March 1964.

Karen's father was a snowmobile dealer, handling the Skidoo brand, and with no school on Good Friday, everyone thought it would be a peachy afternoon to ride on the fresh snow. Karen, her friend Carl, and her mother, Peggy Young, set off on snowmobiles and sometime after 5 p.m. were speeding around frozen Jewel Lake. Luckily they had moved off the lake by the time the quake hit because ice broke in many places, and they might have fallen through and drowned.

The group was on a side hill where Karen had gotten her snowmobile stuck in deep snow. Her mother zoomed up to help her, grabbing hold of the back of Karen's machine and yanking to pull it free. At that precise moment, the earthquake knocked Young to her knees. Karen and Carl were knocked off their feet, too, and actually started laughing. Peggy thought she was having a heart attack. It all seemed funny to Karen and Carl.

"We had no idea," Karen said. "This was all in the very first few seconds. We were sort of rolling around and I got back up and said, 'Mom, the earth is moving.' She realized she was not having a heart attack, and all of a sudden Carl and I are rolling down the hill. She stands up in the middle of the largest recorded earthquake on the North American continent and my mother delivers a lecture. 'You two stop laughing! This is a catastrophe of major proportions!'"

Over the next few minutes they defied the shaking and climbed to their feet. They were in a snowfield atop a gravel pit near some wildly flopping trees.

"That created this really bizarre wind," Karen said. "Where we were there was no crashing, no earth splitting, no falling, no anything, just the sound of our snowmobiles and some buckling of the earth under our feet."

At no time does Karen remember being frightened, only accepting that a power greater than she was in control.

"I remember thinking, 'I guess this is it,'" she said. "I don't know that I knew what 'it' was, but I do remember thinking ... that it probably wasn't going to stop."

When the quake ended, the trio jumped on the snowmobiles and raced for home, a quarter mile away. The house had been built on gravel, not clay, and the U.S. Army Corps of Engineers later told the family that the home was on the most stable part of a bluff. It did not slide into Cook Inlet.

Keith Young, having opted out of snowmobile riding that day, was trimming brush at the side of the house. He decided to break for coffee, and when he went into the kitchen he dropped his heavy waterproof pants to his ankles. Of course, that is when the quake hit.

Suddenly, cups and glasses began falling out of the cabinets, and he tried to hold the doors shut, but his maneuverability was limited because he had rolled down the pant legs and tripped every time he tried to take a step. Awkwardly trying to do several things at once, Young was nearly clunked by a two-and-a-half-foot Italian earthenware vase, one of his wife's favorite pieces.

At that point he stopped trying to save the glassware and thought more about saving himself. He tried to run for the door, but when he began pulling the overpants up they got tangled, so he clutched them with one hand while trying to dodge falling debris and stumbled his way outside.

Young planted himself between his truck and his wife's Volvo, trying to balance between the vehicles and get those darned rain pants pulled back up. The vehicles began shaking and pulling in opposite directions. The ground at the home held, so the only casualties were the dishes and glassware. A sticky conglomeration of jelly, jam, honey, and syrup spilled on the floor and had to be cleaned up with a shovel.

"They were a horrible mess," Karen said.

Once the family realized it was in pretty good shape, Keith Young and a neighbor who worked in construction made their way downtown to see how buildings the neighbor had worked on fared. Karen and her mother drove around picking up cousins and a few other people and brought them back to their home.

The Youngs' house was a gathering place because it had a portable electric generator. Karen's dad was a plumber, so they got their water back on more swiftly than others, too. That night they slept at home, enduring scary aftershocks while listening to radio news reports delivered by Jeannie Chance.

"You stayed close," Karen said. "The radio was always on. You had checked in with your family members. My aunt and uncle lived across the street and had these big forty- or fifty-gallon aquariums, and she lost those. There were fish on the floor and she had cats. That was a mess."

Large swatches of Alaska were a mess. Anchorage was in turmoil. Communities closer to Prince William Sound were hit harder. Tidal waves had been generated. The first reports of disaster transmitted to the Lower 48 were somewhat garbled and frightened relatives living out of state. The power was out. Many had no heat. But the city was not burning down, as had been reported.

It took a little while for the magnitude of the catastrophe to sink in as people scrambled to survive, help friends or strangers, and try to understand what was happening.

The sheer size of Alaska and the vast distances between cities played a role. Juneau, the state capital, is about eight hundred miles from Anchorage. It is part of the southeast region, closer to Canada than most of the state's population.

AT THE TIME of the second-largest earthquake in world history, Governor Bill Egan was taking a nap in Juneau.

Egan's son, sixteen-year-old Dennis, later to become a state senator but then a box boy at Food Land supermarket after school, got off work at 6 p.m. on March 27, 1964. Dennis attended Juneau-Douglas High School and lived in the governor's mansion. His parents didn't want him to become spoiled, so they told him he had to earn his own spending money.

Although Anchorage was 570 miles away, occasionally Dennis Egan picked up a signal from KENI radio in his car. He liked the station's rock and roll music. But instead of enjoying music that day, Egan was hearing news being simulcast from KENI TV. The radio station had been knocked off the air. He heard the television station report that it had just returned to the airwaves following a major earthquake.

"So I went home and woke my dad up," Egan said.

Part of Governor Egan's routine was to work all day at the capitol, come home, take a nap, eat dinner, and then work until midnight. Dennis Egan provided a sense of urgency in waking his father.

"I came in there screaming and yelling, and I said, 'There's been an earthquake!'" He didn't know anything about it yet. All the communication lines were down. Everything was cut."

The governor woke up to a nightmare. Ham radio contact was possible, but Dennis doesn't know exactly what his father used for communication in the early hours.

"Things popped immediately," Dennis said. "My dad got up and went running back to the office, and I didn't see him for a couple of days."

His dad was too busy to come home while dealing with Alaska's greatest natural disaster in modern history.

Bill Egan, after whom a convention center is named in downtown Anchorage, was born October 8, 1914. His father, William E. Egan, died in a snow slide while working at a mine when the future governor was six. The elder Egan had come to Alaska from Newfoundland in 1902, looking for a better life for his family in the town of Douglas, across Gastineau Channel from Juneau.

Clinton "Truck" Egan, the future governor's brother, broke his back as a five-year-old and could never straighten up again. By 1906, the family had relocated from Douglas to Valdez, where Egan grew up and later raised his own family. By the time he was fourteen, Egan was driving a truck for the Alaska Road Commission. There were no rules then about driving ages.

Egan was one of two students in the 1932 Valdez High graduating class. Ironically, considering Egan would become a politician, he was scared of public speaking. He overcame his fear by practicing in front of a mirror.

In 1927, brother Truck bought the Pinzon Bar, a famous Valdez landmark, and Bill sometimes worked there. Truck later became mayor of Valdez and received a mail subscription to the *Congressional Record,* a publication Bill avidly devoured. Bill Egan and school teacher Neva McKittrick were married in January 1941. Dennis was born six years later.

Egan served on the Valdez City Council, in the Alaska Territorial Legislature, and as mayor. He also owned a grocery store in his hometown. After playing an important role in winning statehood for Alaska, Egan was sworn in as Alaska's first elected governor in January 1959, serving until 1966 and then again from 1970 to 1974. He was at the height of his popularity in 1964.

5 Chaos

Mort Henry and his girlfriend, who had come from Fairbanks for a getaway weekend in Anchorage, did not make it to her parents' house on the night of the earthquake. They were not considered missing persons because her family did not know they were in town.

Once they set out to escape the downtown and their quake-damaged hotel, they made their way to the Wendler Junior High emergency shelter, and that's where they stayed. Food and water were being distributed. So many people showed up that the young couple slept on the gymnasium floor.

"It was organized and disorganized," Henry said.

Dozens of people, perhaps hundreds, came by the shelter. Sleeping bags, pillows, and blankets were brought in for those like Henry and his girlfriend who were stuck there with nowhere to go. The only news that people at Wendler received, Henry said, was what arriving families told them. There were no radios at first. The shelter dwellers were in the dark, literally and figuratively. They knew nothing about the tidal waves ravaging other communities and about tsunami warnings for Anchorage.

"The focus was pretty much on right there," Henry said.

The couple stayed at the shelter for four nights. When Henry returned to Fairbanks, he was startled to learn that shake had been strong enough to break glass and cause other minor damage.

"I think we were lucky," he said. "What if we had started walking over there to the restaurant fifteen minutes earlier? We would have been on Fourth Avenue at the time. We could have been inside the Hofbrau."

Photographs of damage of Fourth Avenue show that the restaurant was at the center of the downtown wreckage.

During the early hours after the quake, and in the first days afterward, accurate communication was at a premium. Anchorage was without electricity and phone service. Holes in the streets and roads hampered traffic. There was little direct contact with people in the Lower 48, including those worried about relatives in Alaska.

Edna Kruckewitt, a nurse who lived in Spenard, tried to get word to her stateside family she was okay. She sent a telegram that read, "Everything all right here. Love, Edna." It took four days for the wire to reach Seattle.

After that, Kruckewitt mailed letters to her family. The first, dated March 28, the day after the 9.2-magnitude quake, described the mood of the city.

"Everything confusion here," Kruckewitt wrote to her father. "No one can really tell how bad things are as yet. My house is pretty well shaken up, but it is a well-built building and withstood it well. Have been working at the hospital last night and cleaning up our office today. Some areas are really hit bad. Travel is limited, no one allowed in many areas. Casualty reports are being withheld from the Anchorage population to prevent panic. Entire buildings have disappeared, so the total casualties are unknown.

"Many towns have been completely destroyed. There are huge fissures in all the streets. My floor has a buckle right through the middle—it is a cement floor. I slept through the big aftershock we had at 5:30 a.m. this morning, so it couldn't have been too bad—although the garage across the street collapsed on three cars. Guess I was tired.

"People are rather in a state of shock. Families are separated.

"It was an awful sensation—like nothing that I have experienced before. I knew it was an earthquake when it started, but never imagined one so horrible. The entire house seemed to be bending and twisting, back and forth. I kept thinking it had to stop, it could not possibly get worse, but it kept getting worse. I never have felt nearer to death."

Those who survived the 1964 Alaska earthquake had that commonality of experience. The shaking and rumbling has always been within easy recall. However, many children had a different association with the earthquake. At 5:30 p.m. on Fridays, as moms prepared dinner, Channel 2 aired a popular children's television show called "*Fireball XL5*." A British production, the science fiction/outer space/puppet show featuring Captain Steve Zodiac was riveting entertainment for the elementary-school set.

The show's heroes saved the world a few times in decidedly low-tech ways. The show was for children of a different era but was considered cool and was popular during the early 1960s.

Doug Keil, a ten-year-old *Fireball* fan, was watching his favorite show when the world went crazy around him. The noise, the shaking, and the television show all became intertwined in the minds of him and the other children viewing the screen. Pete Nolan, eight, was looking for his father to complain because the screen went blank. Before he could do so, the power went out. Larry Campbell, eight, was watching *Fireball* with his brother. The noise was freight-train loud, and he could not make out his mother calling to them.

While much of Anchorage—never mind just children—was trying to determine what was going on, government officials were seeking answers, too.

George Sharrock, a district manager for an airline, was mayor of Anchorage, a part-time job. After the ground stopped shaking and damage assessment was called for, Sharrock convened an emergency meeting of military generals, police and fire chiefs, Civil Defense leaders, and rescue workers. They gathered at an Anchorage fire station at 3 a.m. Saturday, more than nine hours after the quake.

The situation was grim. Anchorage was without power, people had been killed downtown. Communication and water supplies were disrupted. There was fear of a tidal wave.

Sharrock was in his car when the ground moved. He thought he had broken an axle. Ironically, he and a friend from Turnagain had just talked about bank erosion on the bluff overlooking Cook Inlet. The bluff was being buffeted by waves, and the man worried that his house was going to be in danger of going off the edge of the seventy-foot bluff.

"I told him one of these days he'd wake up and find his house in the Inlet," Sharrock recounted twenty-five years after the quake. "Little did we know fifteen minutes later it'd come true."

Soon enough, Sharrock realized that he was not having car trouble; he was having land trouble. He watched a raven's unsteady attempts to land on a telephone pole that was moving back and forth and concluded, "Earthquake!"

At the emergency meeting in the middle of the night, as much of the population of Anchorage cowered in unheated homes and looked for guidance how to escape potential tidal waves, Sharrock debated the wisdom of declaring martial law. Anchorage officials consulted with Governor Egan in Juneau by radio. Police were called out for vigilant patrols to watch over half-fallen buildings and homes, especially in Turnagain. The Alaska National Guard was mobilized and patrolled especially hard-hit areas near West High School, in Turnagain, and on Fourth Avenue.

No one felt safe. Aftershocks jangled everyone's nerves. Sharrock said there were six aftershocks with a power of at least 6.0 in the first twenty-four hours. The mayor recalled later that he subsisted on candy bars and sandwiches and about five or six hours sleep during the first seventy-two hours after the earthquake. Sharrock, who died in 2005 at age ninety-four, is remembered as the "earthquake mayor."

With telephones out and it taking hours to get radio stations KENI and KFQD back on the air, authorities communicated using old-fashioned couriers.

Plenty of secondary problems were reported, including a gas line that ruptured and spilled three hundred thousand gallons of fuel into Ship Creek.

Military aid was essential. Soldiers stood guard downtown to protect against looting and to prevent anyone from accidentally causing further collapse of precarious quake-damaged buildings. Military units set up portable toilets, water systems, and four field kitchens.

ON GOOD FRIDAY, people were going about their everyday business. They had no warning, no time to prepare. The trembling was so violent that few moved very far from the spot they were caught in. Only gradually did it become apparent to experts what the strength of the quake was like, where it was centered, and how rare the quake was.

As the world's second-most-powerful earthquake ever recorded, the Alaska quake was number two behind the 9.5 quake registered in Valdivia, Chile in 1960.

Alaska is located on the Pacific Ring of Fire, an earthquake-prone zone, so Alaskans are used to having the coffee in their cups stirred, their deep sleep interrupted, and their overhead light fixtures sway. But when the Good Friday Earthquake struck, it was so imposing that people referred to it as "The Big One" because the power was much more intense.

The earthquake was centered eighty miles east of Anchorage in Prince William Sound under a glaciated peninsula between College Fjord and Unakwik Inlet. The power was so great that one hundred thousand square miles of land was moved and a large swath of land covering six hundred miles between Cape Yakataga and Trinity Island dropped six feet. In the other direction, land rose as much as fifty feet.

The quake inflicted its worst damage to apartment buildings. The Four Seasons was under construction, but nearing completion on March 27, 1964. It was ruined. The fourteen-story Mount McKinley Apartments was destroyed. The 1200 I Street apartments were no longer livable.

ALASKA RAILROAD officials met on Saturday, the day after the quake, to orient themselves to the situation.

"No one had any illusions about what we were going to find out," said Thomas Fuglestad, then the railroad's assistant chief engineer. "We expected the worst, but what we found out even exceeded that."

In many areas, tracks had been obliterated and so had depots. Tracks in the city and leading out of town were tortured into twisted masses of metal, preventing train movements. Elsewhere, tracks were mangled and buildings half destroyed. Avalanches had been loosened by the tremors, sending tons of snow shooting down from the Chugach Mountains between Anchorage and Seward, burying miles of track.

The Anchorage International Airport was shut down. There was damage to runways, but most important, the control tower had collapsed, and the electronic equipment wasn't operating to guide incoming and outgoing jets. Two air-traffic controllers, Bob Daymude and William Taylor, had been working in the tower. The tower was lightly staffed because several other controllers were attending a training session.

Taylor was at the top level of the tower when the earthquake arrived in all its fury. Daymude was one floor below. Both men made a run for it. Daymude reached the ground floor and exited the tower safely. Taylor did not. He was dashing down the stairs when the tower gave way, and he was crushed to death in the rubble. Later, the old tower at Lake Hood was named the William Taylor Memorial Tower and donated to the Palmer Transportation Museum when it was replaced.

The late Dr. William Mills, a well-respected orthopedic surgeon and a frostbite expert, was consulting the last patient of the week in his office when the Good Friday Earthquake struck. Suddenly, he recalled, "the world seemed to be coming apart."

Mills' friend, Dr. Perry Mead, who was soon to learn that two of his sons were killed in the quake, checked out his earthquake-disheveled home and then went to work at Providence Hospital. Injured citizens streamed to Providence and to the Alaska Native Hospital for help.

Mills, who hunkered down at Providence for the next few days, was part of the team of medical personnel working nonstop to treat earthquake casualties.

"We triaged," he said. "We did what had to be done."

It was reported later that Doctor Mead worked heroically on the teams at Providence as two hundred cases were handled during the first twenty-four hours, despite the loss of his sons.

On Fifth Avenue, windows were blown out at Club Paris, the venerable steak institution that was still operating in 2013. At the 515 Club, many bottles of booze were sacrificed to the earthquake. Someone at the Fourth Avenue Theater recalled getting a rain check for a return visit to see *The Sword and the Stone.* With humor, something in short supply that weekend, she recalled that the sword was vibrating in the stone just before the screen went dark.

Norma Lepak Hannon and her husband stopped to shop at the Super S Drug Store on Gambell Street. They were walking their cart down the aisles, throwing in Easter bunnies and other holiday supplies, when the quake disrupted their shopping. Certain the overpowering noise heralded the dropping of an atomic bomb, she threw her coat over her head in anticipation of a flash. Hannon knelt for the duration of the earthquake, only periodically peeking out from her coat to see merchandise flying around.

"I thought maybe this was the way my life would end," Norma recalled in the book, *The Day the Trees Bent to the Ground*, by Janet Boylan and Dolores Roguszka.

When the shaking stopped, she straightened up, looked around, and realized that all of the store clerks were gone. Hannon left the drugstore without the Easter goodies.

The collapse of a concrete wall at the JC Penney store became a powerfully enduring symbol of the Good Friday Earthquake. The store was a year old, a gleaming new building in the old frontier town founded as a tent city in 1915.

Two of the nine people killed downtown died at the store. Huge sections of walls ripped from the side of the building and fell onto parked cars, crushing their roofs. Some of the cars were empty, some occupied. Another complete side of the department store collapsed, as well. Because the store was about to close at 6 p.m., it was not as crowded as it had been earlier in the day.

People were browsing for Easter cards, running an errand to the camera counter, looking over scarves and key chains, examining jewelry, and waiting to pay for their purchases at a cash register. When the quake struck, a young employee left her cash drawer open and ran out into the street.

One of the worst-hit structures downtown, the five-story Penney's not only swayed, it began crumbling. Merchandise was thrown all over the floor. People were knocked off their feet. Instinctively they felt they had to run, to depart the building rather than wait out the quake inside. Whether they realized it or not, sides of the building were peeling off.

Outside, people on the street ran away from the Penney's store. Others grabbed unsuspecting children and yanked them away from the site. Inside, a brother and a sister shopping were going to rush out of the building when adults began screaming, but a postcard stand fell over and pinned the little girl. With her brother's aid, the girl squirmed free and made it out the door just as the entire upper floor walls ripped away from the building and crashed to the sidewalk exactly where they had passed seconds earlier.

Most people got out of Penney's, but a group of four employees rode out the quake huddled in a windowless rear office. Suddenly, a wall tore away, and they were stunned to see outside to Cook Inlet from their perch on the fifth floor. Meanwhile, three boys became stuck in an elevator but were lifted out through an emergency trap door in the ceiling of the elevator.

Blanche Clark walked out of the store about two minutes before the earthquake. She got her car started and slipped into traffic on Fifth Avenue before halting at a red light next to the store. When the walls came tumbling down, concrete landed on the roof of her Chevy Impala and crushed it.

The tonnage did not kill Clark, but she suffered a broken neck and arm and could not move. She was pinned inside the vehicle behind the steering wheel, weakened but alive. She heard a small child say, "Oh, look at the car. I wonder if anyone's in it."

Rescue workers appeared, but it was a difficult and delicate job extricating Clark from the car while not doing her more harm. Strangers worked by hand moving the debris. Progress was slow.

"A bunch of us got around and tried to lift the slab," Fred Kings told the *Anchorage Daily News* years later. "But it was too heavy."

A tow truck was tried next, but it could not move the concrete. In a laborious process, the concrete was jacked up gradually. Piles of rocks were placed underneath to hold it up. Clark was cut out of the car, loaded into an ambulance, and taken to Providence Hospital, where she stayed for five weeks.

There were two deaths at Penney's—a housewife, Mary Louise Rustigan, and a high school student, Lee Styer, both killed by falling concrete.

The earthquake shook Anchorage to its core. Residents were frightened. Families lost homes. People were missing, and no one knew if they were alive or dead. The first hours after the quake were confusing, and the threat of a tidal wave hung over the city.

Residents in neighborhoods with little damage might not have known of the tidal wave threat without battery-powered radios or other communications. But those listening to the first radio reports of the disaster were horrified to hear that a tsunami might create new jeopardy.

Evelyn Rush, who was with her family assessing damage at her home, reported that one radio broadcast at 10 p.m. provided an urgent warning. "This is a confirmed report!" she remembered hearing. "A tidal wave is approaching Anchorage."

Citizens living near Cook Inlet were urged to evacuate the area. Before Rush loaded her children into the car, a fire department vehicle came through her neighborhood, reinforcing the message, which essentially was to run for the hills. Believing she had only fifteen minutes to depart, Rush found herself running out of time before she sped to a possible safe zone at the other end of the city. She sat awake in her car until 4 a.m., dreading a tidal wave, but then concluded that none was coming. None did.

Aftershocks created fresh terror through the night and into the next day, but KENI was back on the air trying to soothe people as well as inform them.

Mike Janecek returned to work downtown the next day and manned the company phones. To get to the station, he had to be accompanied by and vouched for by the news director because the military was turning people away from the worst-hit areas. They were protecting against looters and did not want curiosity-seekers to get in the way of rescue and recovery efforts.

Considering that most phone lines were supposed to be out, Janecek was kept quite busy. People would call the station and tell him that someone was unaccounted for and ask that the name be broadcast.

People who had gone skiing for the weekend at Mount Alyeska were trapped about thirty-five miles away from the city due to closure of the Seward Highway. Paul Crews Jr., a well-known Alaska skier and later the ski coach at the University of Alaska Anchorage, drove as far as he could toward Alyeska. Then, on foot and on skis, he made his way over avalanches covering the road to gather names of people stuck at the ski lodge. The names were broadcast on the radio.

"We were fielding phone calls from all over the world," Janecek said. "That was all I did. Very few phones were working in Anchorage."

The Salvation Army stopped by the broadcast station and fed the workers. Janecek said dinner was chicken cordon bleu, a better-quality meal than Army rations. When Janecek went outside to help carry in the food, he couldn't take his eyes off the military guards marching back and forth.

"The city was silent except for in the distance there was a dog howling," Janecek said. "It was a really eerie feeling."

School was out and not about to be called back into session, so Janecek worked day and night at the station, making $1.75 an hour. His small role was trying to help by being a conduit of information. Janecek did not get overtime and for years he joked with Al Bramstedt Jr. that he was owed big bucks from 1964.

"It was an experience beyond imagination," Janecek.

That was an apt description for all Anchorage residents who lived through the Good Friday Earthquake.

6

Anchorage Copes

Bob Reeve, the famed Alaska Bush pilot, who played a role in the founding of the Anchorage Glacier Pilots Alaska Baseball League team named after him, was sitting in a bar in the late afternoon on Good Friday.

His choice of site for his sixty-second birthday celebration was the Petroleum Club on the fifteenth story of the Anchorage Westward Hotel. When the 9.2 earthquake hit, it made Reeve wish he was in the air in one of his planes because that day he went flying in a totally unfamiliar manner.

Reeve was heaved onto the floor of the bar, one of the highest points in Anchorage, and then tossed back and forth as the furnishings around him were buffeted and broken. Liquor bottles and glasses shattered. Tables and chairs were thrown about.

It was good fortune that the hotel, and especially its higher floors, did not suffer more damage. Once the shaking stopped, Reeve knew that the party was over. Aided by a hotel employee, he said he made a slow, careful descent via fire escape, a step-by-step retreat that took about half an hour. Those drinking and celebrating with Reeve were on the ground much swifter than he.

"They made it down a lot faster," Reeve told an interviewer. "After all, they're young and spry."

Reeve, who died in 1980, had his share of adventures in World War I, as a barnstorming pilot, and delivered mail in South America for Pan American World Airways. Later he started fresh in Alaska with two dollars in his pocket after stowing away on a steamship headed north. He developed a reputation as a daring-do, bring 'em-back-alive pilot in the early 1930s, when during the course of two separate medical emergency rescues, he was forced to land on frozen water because of weather. Well before the earthquake, Reeve had established he had nerves of steel.

Reeve may or may not have realized that some people stayed up high on the fifteenth floor longer than he did during the quake. Bartender Cliff Cooper said he was lucky he didn't get cut by flying glass.

"The room was moving so much I had the feeling it was headed for the ground," Cooper said. "The building was rocking. Plaster fell. As we crawled and ran, the hallways buckled and rose as much as forty-five degrees."

Despite its height, the 175-room Westward held together in good shape and was reopened quickly.

"This is the tallest building that has ever withstood an earthquake of an intensity as great as we had, or anywhere near to it," Seattle-based architect Bob Durham reported to the Anchorage Hotel Corporation.

PUBLICATION OF the *Anchorage Times* had been interrupted by the earthquake, but by Sunday, March 29, reporters were on the ground and the newspaper was on the presses, attempting to fill the information void only intermittently handled by radio stations.

Monday's paper hit the streets bearing a large front-page headline:

104 FEARED DEAD IN QUAKE AREA
City Begins Major Reconstruction

To Bob Atwood, it was inconceivable that all he had sought and fought for so hard and so long—Alaska statehood and economic growth—could be jeopardized by an earthquake. It was difficult to sugarcoat its severity, the damage it did, and the harsh reality of crumpled buildings, but Atwood was nothing if not a booster, and his rah-rah attitude transferred to the mood of much of the coverage.

Like many of Alaska's political leaders, Atwood immediately began throwing out ideas and drawing up plans for a rejuvenated Anchorage and Alaska. Almost before everyone in a collapsed building was evacuated, he was discussing replacement structures.

A certain human touch, so important at a time of mass public crisis, seemed to be missing from the coverage in favor of the attitude of "We'll make it all better." Residents were still assimilating deaths, destruction, and loss and just beginning to grieve, yet on Tuesday, March 31, the headline, equally as large in type as the one the day before, read:

REBUILDING BEGINS IN ALASKA

That was a preposterous outlook. People whose homes were wrecked had not even reclaimed their available possessions. The death toll was incomplete. People were missing. It was premature to proclaim that rebuilding was underway. The ground had stopped shaking, but people hadn't.

Residents of Seward, Valdez, and Kodak, in particular, might have found Atwood's approach offensive—if they knew about it. They were isolated, facing local destruction of a scale none had imagined. They were still more in survival mode.

The *Times*' positive headlines were accompanied by grimmer, more realistic ones:

22 DEAD, 50 MISSING ON KODIAK ISLAND

WHITTIER'S GASOLINE TANKS STILL BLAZING

COMMAND RELEASES CASUALTY FIGURES
More Than 30 Persons Are on Missing List

Valdez, Seward, Whittier, and Kodiak seemed to be hit harder even than Anchorage, though how badly damaged they were was not immediately clear. The smaller communities were more condensed, losing entire neighborhoods and commercial districts, with more direct tsunami impact.

The *Anchorage Daily News,* the *Times'* smaller competitor, gave more voice to individuals and their quake experiences in its March 30 edition. The *Daily News* also missed Saturday publication and returned Sunday.

Elizabeth Trigge was sitting in her car waiting her turn at a drive-in banking window near JC Penney when the world seemed to shake apart around her.

"Penney's looked like a cake crumbling," she said. "The street was heaving and the wires dancing. I looked down toward Penney's—stuff was falling from it to the street."

Trigge abandoned her car, grabbed onto a post, and then ran for her life as Fourth Avenue began breaking up.

Falling debris trapped many people in offices and apartments. One man was pinned down in a bathroom of a downtown apartment building. The man was feared to have two broken arms and was experiencing chest pains.

"First we used axes," said rescuer Tony Parisi. "Then I ran and got a power saw from the Alaska Railroad."

Clark Phillips, a customer at the Tip Top Café on Spenard Road, told a strange story of he and the owners and an unidentified customer being present in the café when the earthquake hit hard, shaking the building violently. They all fled, except the unidentified man.

"This other guy sat right there at the counter and continued reading his paper and drinking his tea through the whole thing," an incredulous Phillips said. "He didn't say a word. He didn't even look up."

Marie Kroon was home alone in Turnagain when the quake battered her house, breaking it apart. Kroon exited through a gigantic hole where an outside wall had stood, stunned by the loss. She told people she was glad that her husband and sons weren't home at the time—and then learned they were shopping at the JC Penney store.

Later, Karl Kroon explained how he and his thirteen-year-old twin sons made a run for it from Penney's as they feared the building would tumble down around them.

"The boys were trying on suits, and the lady was fitting the pants when the building began to shake," he said. "The corner of the building started going, and we started running for the escalator. There were about fifty people scattered throughout the floor. The whole thing started swaying."

The escalator was not working, so they switched to the stairs and descended to the first floor. One of the Kroon boys stumbled over a mannequin, thinking it was a body and said, "Daddy, pick him up."

Kroon said people behaved well, not panicking or screaming, but moving fast, he was stunned to see the damage as they ran.

"The wall just fell out and we could see daylight out on Fifth Avenue," he said.

Penney's was the site of first destination of the Anchorage Fire Department. The fire department's clocks stopped at 5:36 p.m. as power and communications were lost to almost all stations. It took four minutes, however, for back-up generators to fire up and be put to use.

"A number of people on the street or in cars at this location were buried under the debris," the department's report on the Penney's response read. "R-2 (unit) was activated and responded from Station 3. Bodies were found immediately and one severely injured woman was removed from an automobile and taken to Providence Hospital by R-2. 151 responded to 4th and B to the Denali Theater where it was found the building had sunk below the 2nd story. No injured were immediately apparent at this location."

Much of the fire department's role in Anchorage consisted of rescues and emergency transporting of people hit by debris. Anchorage was spared the type of major fires that often follow earthquakes when natural gas lines explode.

The lack of fires was a blessing because the city's primary water supply and mains had been shut down by the earthquake. A 6:10 p.m. log-in reported, "There was a complete water failure in the downtown area and all fire hydrants were out of service. The only immediately available water was in the booster tanks of our engine companies and the 1,100 gallons of water in Tank 1."

By 6 p.m., the fire department was on its way to Turnagain after ascertaining that Government Hill Grade School "had fallen over the bluff and been destroyed."

A Turnagain resident, Del Ingram, thought the quake was going to be like many others he and his family experienced, a sharp jolt maybe, but ending quickly. This one didn't end quite so fast.

"The garage fell off first," Ingram said. "We tried to get out the back, but nothing was there. We ran out the front. Each step we made was dropping down under us. We just kept going. There was no telling where it would stop."

The Ingram home was thrown about two hundred yards down the bluff above Cook Inlet.

Even before fires were doused in the smaller southcentral Alaska communities of Seward and Valdez, the Anchorage City Council met in emergency session on Monday, three days after the quake, to pass ten resolutions requesting financial aid from the federal government in order to cope with the catastrophe. The requests were aimed at obtaining financing to fix infrastructure like streets, sewers, water systems, telephone wires, and oil storage tanks.

Governor Egan set up an emergency headquarters in Anchorage, where he met with his cabinet and fielded offers of help from generous contributors around the country.

The governor asked that donations to be sent to his attention at his Civil Defense headquarters, a trailer office, at Fifth Avenue and Juneau Street, where he relocated for meetings with military, law enforcement, and elected officials. Alaskans were going to have to depend on the largesse of the federal government for help in repairing the massive damage.

Parts of downtown and Turnagain were sealed off by authorities, and traffic was routed around the most affected areas.

"We will prohibit public access to the worst areas," said Anchorage city manager Robert Oldland, "pending such time as inspections can be completed and buildings declared structurally safe. The banks will be inspected first."

"Do not enter" decrees were announced for the blocks of Third and Fourth Avenues at A, B, C, and D Streets. Exceptions were made for access to Hewitt's Drugs and Ellen's Jewelers.

Sympathetic to the new state, President Lyndon Baines Johnson was in Johnson City, Texas, his home state, for a five-day Easter vacation during the aftermath of the quake.

After speaking to Johnson by phone, Governor Egan said the president "expressed his utmost concern" about Alaska, and promised he would do all he could to help the people of the state.

Johnson quickly announced executive orders for financial and emergency assistance. Then he cut his trip short, returning to Washington, D.C., in order to oversee the operation. Edward McDermott, a special envoy to Johnson, was sent to Alaska to discuss the problems firsthand with Alaskan leaders.

Egan was also obtaining reports funneled to him from the field. Three days after the quake, he reported that the death toll was higher. In all, stretching to California, where the resulting tidal wave proved fatal to some, there were 131 deaths ascribed to the Great Alaska Earthquake.

"Many have said it is nothing short of a miracle that the casualties were not far, far greater," Egan said in a speech. "I can think of no finer tribute to those that are gone—many of whom I knew personally and will deeply miss—I can think of no finer memorial than to work on another miracle.

"A miracle of will and energy and faith that will make this great land of ours better than the one we knew before. From what I have seen accomplished during the past seventy-five hours I have no doubt that we will succeed. No man could ever have had more pride than I do tonight at being governor of this great state."

There had been some good news—other than the lack of more casualties and property damaged. At Anchorage's Providence Hospital, where many of the injured were taken, Nancy West had been working the switchboard that Friday afternoon, and it fell to her to make a rare announcement over the public address system: "Please remain calm. Stay in your room. The hospital will stand."

It did.

The 155-bed hospital was new at the time, having been constructed at a cost of $6 million. Providence was the medical emergency hub for the community. After initial swaying, and a loss of water, the structure endured. Then doctors and nurses took over caring for those harmed.

In other news that could be considered upbeat, the *Anchorage Daily News* reported that at least seven babies were born in the city on quake day, six of them on schedule. The seventh was slightly early, perhaps due to the unexpected shaking.

Although there were no immediate food shortages in Anchorage, emergency supplies were donated in the Lower 48 and flown north on commercial and military airplanes. Supermarkets had been damaged, and some feared of looting, but the fears quickly subsided. On April 1, representatives of the grocery industry took out a full-page ad in the *Anchorage Times,* thanking the public for "patience and cooperation during the present emergency."

The self-described "Anchorage retail food industry" said it would cooperate with Civil Defense and other authorities, try to serve customers the best it could, keep prices "normal except actual cost of emergency transportation of certain special items," and "to adhere absolutely to our policy of discouraging scare buying or food hoarding … with your continued cooperation we can all hope that food rationing will not become necessary."

That need for rationing never materialized.

Still, the comfort zone of Anchorage residents varied. The night of the earthquake virtually all electricity, telephone, and water supplies were shut down. Gradually, neighborhood by neighborhood the services were restored. Many people slept huddled in blankets next to fires set in still-serviceable fireplaces the first night. On March 27, the weather remained winter-like, with a light dusting of snow on the ground and periodic snow flurries in Anchorage.

"For the first twenty-four hours," Peggy Bensen said, "you filled every bucket and container with water and didn't worry about it. It wasn't that cold and in our trailer we had oil heat, so we could keep the heat on part of the time. I could turn the propane gas range on, and warm up the kitchen that way, too.

"We were very well off. We had lots of food in our cupboards and the neighbors could come over and get whatever they wanted. People ran out of milk and I had lots of canned milk. We were prepared. We just pretty much stayed home for a week. It could have been worse."

The U.S. Army provided water, cholera shots, and other emergency supplies at Wendler Junior High. Hundreds came to the shelter.

AN ALASKA RAILROAD employee, Dick Mackey, reported to the federally owned railroad's Anchorage headquarters a few days after the quake. The railroad was decimated. In Portage, about forty-five miles from Anchorage, the tracks were destroyed, the highway was a mess, and the ground had sunk about seven feet. The highway bridges were out, but the railroad bridge still stood.

Mackey said he watched soldiers from Fort Richardson walk along the road with poles to probe cracks of the highway. They were making sure the road wasn't about to give way. Portage had been almost completely destroyed. Survivors were staying at Diamond Jim's Bar in Indian. Escorted by the military, Mackey drove the first supply truck to people taking refuge in the bar.

At one point, Mackey had to turn the truck around and back it over a railroad trestle.

"The ground had sunk on the Portage side of the railroad bridge and they had put some ladders up there," he said. "Soldiers unloaded all the supplies and got them over to Diamond Jim's Bar.

"It was a long day," Mackey said.

The occasionally funny story made the rounds amid the hardships and sadness of lives lost. Karen Sobolesky, who had been snowmobiling with her mother, said the family knew someone who was working at the soda fountain at Penney's. One of the friend's co-workers began screaming, terrified that she was dying because of all the blood pouring down her legs. Sobolesky's acquaintance ran over, felt the "blood," and realized her partner's legs were coated in strawberry jam.

Young people like Sobolesky and John Schandelmeier were out of school for a week or more. No one seemed to mind the unscheduled time off from school.

"I was just crushed," he joked, about being out of school. "I always did well in school, but I never liked it."

One thing he did with his free time was take long walks in the woods. He was a hunter and trapper already, and he was at home in forest land. He went exploring to see what type of mark the earthquake made away from buildings.

"There were big cracks in the ground," Schandelmeier said of the area south of Anchorage, which is now developed, but was not at the time. "We had a trail back there and I remember a big crack that would be just short of where the new Seward Highway is now. It went for a long ways. I followed it. I remember walking along that crack for miles."

When school reopened, all high school classes were held in double shifts at East High School because West High had been heavily damaged.

There was a brief concern that Alaskans might flee the state, either temporarily or permanently, because they feared another major earthquake repeat.

However, when commercial aircraft companies resumed their schedules, passenger traffic to the Lower 48 was even less than normal.

People came to The Last Frontier because it represented living on the edge, was a bounty of wilderness, and differed from other states. Even the second-largest earthquake in world history couldn't drive them away.

"The real Alaskan will stay here no matter what," said Joan Copeland, a waitress at the Westward Hotel.

Grocery-store employee W.W. Kelly, who suffered a hand injury in the quake at the Market Basket where he worked, said, "I never believed in running from trouble."

That type of resolve was common.

"This is our home," said Anchorage resident Donna Vincent.

Many predicted that Anchorage and other hard-hit communities would be rebuilt completely and would continue to grow. That has proven to be true.

Many business owners pledged to rebuild. So did JC Penney. The national chain department store was rebuilt and refurbished, and it was very much an anchor store in downtown Anchorage some fifty years after the Good Friday Earthquake.

7

Seward under Siege

Dan Seavey and his wife Shirley, both teachers, left central Minnesota behind when they moved to Alaska in August 1963, in time for the new school year. They had teaching jobs at Seward High School.

On the day that would cause terror and heartache throughout their new hometown, Shirley, a member of the community choir, attended a practice session in advance of a scheduled performance that night. Dan took advantage of the fresh snow and mild temperatures to go snowshoeing with a friend in the Snow River Valley about fourteen miles north of town.

Although it was overcast, Seavey said it was "a nice promise-of-spring" day in the backcountry. Seavey and his partner returned to town in late afternoon. If they had lingered any longer, they might have been trapped because the earthquake closed bridges into the city.

A decade later Seavey would play a role in developing the Iditarod Trail Sled Dog Race with Joe Redington Sr. and other friends. He placed high in the first of the 1,000-mile Iditarod races across Alaska in 1973, and his son Mitch later became a race champion. On this day in 1964, Seavey had borrowed snowshoes from Tom Johnson, another Redington pal and teacher who later also served as an Iditarod organizer.

Seavey stopped at Johnson's house to return the snowshoes before heading for his home a few hundred yards away. Shirley had come home to prepare an early dinner so she could leave for the choir concert. Dan lingered at Johnson's home for a while, just shooting the breeze.

As Seavey stood up to leave, Johnson pointed to a pint jar of fifty-cent pieces and said, "I'll give these to you if you can guess exactly how much is in there." Seavey took a calculated guess and amazingly guessed the right answer.

"I forget what the number was," Seavey said later, "but it was right. He was going through all kinds of fits because he was going to have to give me that jar of fifty-cent pieces."

Just then, the rumbling was heard. The two men looked at one another, puzzled.

"We were wondering, 'What in the world is that?'" Seavey said. "About that time the ground started to roll. I had never been in an earthquake, but in unison we yelled, "Earthquake!"

Looking out the front door, the men watched cars bouncing in slow motion and electric poles swaying back and forth. They didn't know whether to run outside or stay inside, so they pretty much stayed rooted to an area near the front door, managing to remain on their feet even as the ground rolled.

"It was like a big snake under the ground, kind of moving along," Seavey said. "It was like waves and there was the noise, that big rumbling. We ran out in Tom's front yard after a while and about then things began to explode. The Standard Oil dock and the tanks on the dock started to puff up and you could see flames and explosions."

The docks were only seven blocks away, so the explosions were easily heard, and the flames shot high enough into the air to be seen for miles.

Seavey watched spellbound as the freight docks and Standard Oil began going up in flames. A giant crane stood perhaps six stories high on the dock.

"I could see it was dancing around," Seavey said. "It was doing that elephant dance. Then Standard Oil tanks went and my attention was distracted. It was diverted and by the time I looked back the crane was gone. It had fallen, tipped into the bay."

As soon as the earth stopped shaking, Seavey sped for home. Shirley was there with their three children under five years old. A neighbor was on the porch informing her that there had been an earthquake. The kids were crying and carrying on.

"They were like the rest of us," Seavey said. "They didn't know what was happening."

Remarkably, despite living so close to the community's major earthquake damage, there was no damage to the Seavey home; only a small amount of glassware broke.

"What we lost, they say, was a half a block, or even in some places, a whole block all along the shore line of Seward," Seavey said. "That's where the industrial base was, the railroad, the Coast Guard, the fisheries, the small-boat harbor, all of that stuff."

The violent earthquake shook Seward hard and was quickly followed by a powerful and deadly tsunami with huge waves rushing in from Resurrection Bay. The community's perch on the dazzling bay had always been an asset, economically and recreationally. Goods and materials moved over the docks. Pleasure boats and fishing craft dotted the surface on pleasant days.

This day, the scenic bay became the enemy of the city nestled beneath famed Mount Marathon, the 3,022-foot hunk of rock looming above downtown that has been the site of an annual July 4 mountain run since 1915.

"There was more than one tidal wave, and that was frightening," Seavey said.

Long after the ground stopped shaking and the water receded, the fires at the docks kept roaring.

The earthquake unleashed its destructive power only two days after Seward was named one of the nation's ten All-American Cities by *Look* magazine and the National Municipal League. The *Anchorage Times* had produced a special twelve-page section about the community's economic development and bright future.

Now Seward was in shambles. Many homes and commercial buildings had been wrecked, and thirteen members of the community had lost their lives. The fishing fleet suffered $1.8 million in damages, and at least eighty-three homes were damaged beyond repair. Altogether, estimates of damage reached $50 million dollars.

A FEW DAYS AFTER THE QUAKE, Chuck Lastufka, an Anchorage-based pharmaceutical sales representative for Pfizer, took to the air in a DC-3 to find out if drugstores in the region had sufficient stocks of medicine for their patients who took ongoing maintenance prescriptions for diabetes, high blood pressure, and other chronic illnesses.

All of Alaska was Lastufka's territory. On Thursday, he had overnighted in Homer and had visited Kodiak on Friday. He returned to Anchorage late afternoon that day and had been sitting at his kitchen table for fifteen minutes with his eight-months'-pregnant wife, Francine, staring at what he estimated was ninety minutes' worth of paperwork to be dealt with before he could declare the holiday weekend underway.

Lastufka was writing at the table when—Boom!—he heard a loud bang. The initial impact was so strong at his home on a street off of Tudor Road in midtown Anchorage that his chair went over backwards with him in it.

The refrigerator fell over. A sixteen-place china set burst free of the cabinets and broke into pieces on the floor.

"One cup was left," Lastufka said.

He scrambled to his feet. Francine was in the living room, and he shouted, "You've got to get out of here." To do so, they had to make their way over and around Chuck's moose heads and other big-game mounts that had been knocked off the wall in the entry way.

The couple struggled over the shaking ground and reached their station wagon but couldn't get into it. Francine leaned across the front of the hood to balance herself. They watched the driveway crack open, with holes yawning wide and then closing. The spruce trees swayed and bent, their tops nearly touching the ground before straightening up again. They rode out the long earthquake with the car for support and were so focused on staying on their feet that they almost forgot to be scared.

"It felt like an eternity," Lastufka said.

Eeriness descended with the end of the quake.

"There was a big silence," Lastufka said. "That's what you noticed. All of this dust was swirling in the air. But it had become very quiet."

Not for long. Neighbors poured into the street. Head counts were taken. Everyone seemed present and accounted for and unhurt. But Lastufka and others recognized that there had to be injuries and deaths elsewhere.

Back in the house, Lastufka got his battery-powered radio going and heard the first reports of devastation in Turnagain. Later, friends appeared at the Lastufkas' door, carrying sleeping bags and looking for shelter. With the power out, Lastufka ignited a fire in the fireplace and lit a Coleman stove. By evening, twenty-five people were staying at the Lastufka home. Fortunately, Lastufka had a freezer full of moose meat, which his group cooked on Coleman stoves.

Curious, Lastufka drove downtown Saturday, working his way past broken and blocked-off streets. The military presence was strong. Word spread that soldiers would shoot looters on sight. The scene in Anchorage made Lastufka wonder what had happened in the smaller, outlying communities he served.

Lastufka felt a responsibility to his customers. He could not get detailed reports from Seward, Kodiak, Valdez, and elsewhere. What news did filter out was grim. Likely those suffering from long-term illnesses requiring daily medication had lost their medicine in the earthquake or tidal waves, and some of the drugstores had been shuttered. Because many clients had debilitating illnesses that were controlled by medication, Lastufka knew it could be dangerous for them to have treatment routines interrupted.

Lastufka wrangled a seat on a DC-3 carrying supplies. His first stop was Seward. A few days after the earthquake, no one had been able to put out the fires.

"The oil tankers were still burning," Lastufka said.

The effects of the tsunami made a strong impression on Lastufka. Boats tied up at the docks in the small-boat harbor had been lifted from their moorings, set loose on the water, and carried inland. The tidal wave's force scattered the boats throughout town and as far as the edge of nearby mountains.

"I'll always remember what that looked like," Lastufka said.

Like most communities caught up in the catastrophe, the most serious impact on Seward was selective. The business district was virtually destroyed. Homes located closest to the water were most seriously damaged. Other parts of town were barely harmed, and people throughout the community, newcomers and old-timers alike, gave thanks that more people hadn't been killed. A public dedication of the new fire station scheduled Sunday took place, but it was less a celebration of the new facility and equipment and more a gathering for prayer and an opportunity to boost the somber mood of survivors.

FLIP FOLDAGER, a life-long resident of the Seward area and an acclaimed Mountain Marathon runner, was a first grader when the quake almost wiped his town off the map. Just as nearly every American of a certain age recalls where he or she was when President John F. Kennedy was assassinated in November 1963, everyone of a minimum age from Seward remembers where he or she was on March 27, 1964.

Foldager, then six, lived with his family in an apartment above the Estes Brothers Store in Moose Pass, about thirty miles from downtown Seward. A bunch of children had gathered there to watch *Fireball XL5*. The show was the perfect kid distracter, engrossing youngsters while mom cooked dinner.

"We were sitting there watching," said Foldager, who as an adult has been employed by the city of Seward as a harbor worker for two decades. "All of a sudden everything started shaking, of course, shaking like hell. The older Estes boys started getting all of us kids out. You had to go down a pretty long wooden stairway from the second story. I can remember standing in the yard after they got us out and one of the older boys, one of the last ones going down the stairway, was still on it as it collapsed. I remember him falling down to the ground. He didn't get badly hurt or anything, though."

As the kids huddled together, the earth ripped open nearby with a rumble, leaving a crack in the ground that swallowed two legs of a swing set. Those pieces were crushed when the ground closed. A nearby parked car was bouncing up and down.

Out on the Seward Highway, "the pavement was crackling and buckling," Foldager said.

Most of the children were too young to understand exactly what was going on, and with the temporary news blackout not very many adults could explain it.

"I do remember how it just kept going and going and going," Foldager said. "It was a very long earthquake. I remember watching the power lines swing and things like that. The power poles swung back and forth pretty radically. I don't remember being scared. We weren't really scared. It was more like, 'Wow! What's this?' I was young enough to think like that, and maybe I was scared and just don't remember."

On the other side of the house was Trail Lake. The water level abruptly dropped and almost disappeared completely. The children could see fish flopping around in the mud. It didn't take any adult lectures for them to realize something was badly askew.

Foldager's younger brother Mike and sister Tanya were elsewhere on the Kenai Peninsula, visiting their grandmother. Bridges had collapsed and it was difficult to get back to Seward by road. Several days passed before Foldager reunited with his siblings in Moose Pass.

"They somehow got to the other end on the lake and their final part of the journey home was in a skiff," he said.

Meanwhile, Foldager's father, Jack, was in Seattle, anxiously awaiting a commercial flight back to Alaska. The elder Foldager could hardly believe his eyes as he watched television footage shot from airplanes flying over Anchorage and other communities, including Seward, displaying widespread destruction. The footage from Seward was particularly dramatic because of the fires raging at the Standard Oil Company facility.

"The word was—and my dad said that's what it looked like on television—that Seward was gone," the younger Foldager said.

A ham radio operator in Seward was issuing reports to anyone tuned in around the world on the right frequency. From his vantage point, there was a lot of bad news. He did not know how many people had been killed or injured but did put out the news about the tidal wave and fires and that there had been considerable damage along the water.

Jack Foldager's family was okay, but Seward was in bad shape.

The rising flames were bright orange, but they gave off black soot, as well, from what was burning, and this was all against the background of snow-topped mountains. The fires would have been worse if the wind had been stronger and blowing toward residential areas.

"But it was absolutely dead calm and the fire didn't get blowing all over town," Dan Seavey said.

Those caught up in the tsunami took the brunt of the earthquake's immediate aftermath. Many weathered the shaking and cracking open of the ground but were in the path of a wall of water rising from Resurrection Bay. The first wave rolled into Seward between twenty minutes and a half an hour after the quake, depending on who was timing it. Estimated at thirty to forty feet high, the tidal wave blasted into the community, moving at an estimated one hundred miles per hour, with one serious by-product being the explosion of forty railroad cars filled with oil.

The first tidal wave that rushed into Seward was generated by landslides in Resurrection Bay. It was followed by waves with more power, stirred up by the quake under Prince William Sound near the epicenter.

"The water just emptied completely out of the Bay," said Seward resident Emmitt Hill in a video program called *Oceans of Fury: Tsunamis in Alaska,* produced by the Alaska Sea Grant program.

But the water was coming back. Patricia Williams, who was with her mother, recalled hearing urgent warnings of "Get out! Get out! Tsunami coming! Tidal wave coming! Get to high ground!"

Long-time Seward resident Bob Eads and his brother saw the water coming.

"The ocean was boiling," Eads said.

They tried to make a getaway in their pickup truck, but the water was moving too fast to escape.

"It overtook us," Eads said. "We were under about fifteen feet of water. I thought it was just about the end."

But the water passed over them, and they survived.

When the water receded at one point, after the initial big tsunami, resident Al Burch said it was startling to see the damage.

"All the docks had completely disappeared," he said. "There were a few stumps sticking up there. Basically, for us, Seward was gone."

Indeed, Patricia Williams turned to her mother and said, "Well, mom, there goes your All-American City."

There, too, could have gone the people who had made Seward an All-American City. Death, destruction, demoralization—all could have contributed to a mass exodus of people fleeing the little city overwhelmed by the multifaceted disaster.

After the initial quake and tidal wave, Dan Seavey said he and other male teachers were deputized to patrol the streets to guard against looters.

"The town was completely black except for the fire along the beaches," he said. "I think we did that a couple of nights and then the National Guard showed up. There was no looting that I was aware of. We broke it up into eight-hour shifts and we had our own areas to cover. After that we closed school down, turned in our grade books, dismissed the kids and families started leaving like rats from the ship. They wanted out of this place."

Actually, within a few days four hundred Army troops from Fort Richardson in Anchorage were dispatched to Seward to help in the emergency. Soldiers from the 4th Battalion, 23rd Infantry, and the 562nd Engineer Company of the 172nd Infantry Brigade helped man a command center with Seward Mayor Perry Stockton and patrolled the streets.

Bill Hamilton, an Army private, wrote an account of what he saw. He said more damage was done by far due to tidal waves than the earthquake, and by fire near the docks. Some soldiers were already stationed in Seward. Private Pepe Fuchs reported to Hamilton that moments before the earthquake he and others were watching a train through a mess-hall window.

"We see the train, the whole train with brand new cars on it, and it gets tossed into the water like nothing," Fuchs said. "The whole train just gets flipped into the water and the bridge it was on was gone and the tracks are popping like rubber bands."

School closed for about two weeks. An Army recreation camp and schools were used for temporary housing.

"Then schools reopened to cut down on the exodus," Seavey said.

When life calmed down a little, Dan and Shirley talked about returning to Minnesota.

"It was devastating," he said of the scale of disaster.

However, the setback made the Seaveys more determined to stay in Seward.

"The town was down," Dan Seavey said. "Everybody was leaving and I thought, 'Holy, cow!' To be honest, our plan was to come up to Alaska and get all of the hunting and fishing out of our system and then go back to the Midwest to teach or something. I think it was the earthquake that made us dig in. We're going to stay here."

Nearly half a century later, the Seaveys were still well entrenched in Seward.

8

A Family's Tale of Terror

Linda McRae woke up excited. Good Friday was the day before the Seward High School junior prom, and the fifteen-year-old sophomore had a date with junior Steve MacSwain, whom she later married.

There was no school, so Linda and some friends went shopping downtown. Not that downtown Seward was a very expansive place. Her father owned a hardware store-lumberyard near the old city dock. The city's population in 1964 was 1,891. Seward had been founded in 1903 with the purpose of providing rail transportation inland so mining companies could extract coal. It also marked the beginning of the Iditarod Trail that led miners and trappers into interior Alaska, where they settled several gold-rush communities.

The announcement of becoming an All-American City led to plans for an upcoming two-day Seward festival.

"Seward is busting its buttons today," said City Manager James Harrison.

Soon after, Seward was busting and burning with no celebration on tap for quite some time.

Linda said her family, the McRaes, were old-fashioned like the television couple Ozzie and Harriet. They always ate dinner together and thus Linda, her two brothers, and parents were gathered around the dining room table at 5:36 p.m. when the house began vibrating.

Linda's oldest brother, Doug McRae, was there by chance. The father of a new baby, he lived with his wife three blocks away. When it became apparent this was no passing shake, Doug jumped to his feet and ran out of the house heading for home.

The violence of the quake was stunning. It was as if the kitchen exploded. Cupboards opened and dishes and glasses began flying and crashing onto the floor. At first, Linda thought the powerful noise and shaking was the noisy boiler in the nearby grade school.

"At first, I thought, 'Oh, that must be going on at the school,'" she said. "But it got stronger and stronger and stronger and didn't stop. It got very quiet in the house. Nobody was saying anything."

Linda's father, Scott, told everyone to pack up everything they might need and get into the car. Everybody was worried about Doug and his family, who lived near the small boat harbor.

A "self-centered" teenager, as she described herself, Linda said she jammed all types of things into a bag, including her new prom dress. The apricot-colored outfit had a Jackie Kennedy bell skirt with a bow and the high empire waist like the First Lady had worn and that was the rage among the fashion conscious.

"That day I had just bought a new angora jacket that went with," she said. "It was really to die for. I gathered all of it up and walked downstairs."

The family, including the pet cocker spaniel, jammed into the Corvair. This was the beginning of a difficult exodus. A traffic jam formed as residents sought to escape from the city. All the traffic was headed the same way, but the McRaes turned off toward a subdivision close to the bay, where they thought the Lemas family, Doug's in-laws, might need help.

They reached the subdivision, a small one with perhaps ten houses, but the Lemas home was empty. As they looked back toward town, they saw oil storage tanks and rail tank cars engulfed in flames. The fire grew bigger and brighter as dusk settled in.

The Seward waterfront and docks had slid into Resurrection Bay. The Standard Oil Company dock ruptured and oil was spreading everywhere. The McRaes were stunned. This was their home, their community, and their source of livelihood. All of it was in jeopardy. Linda's father and brother headed downtown to volunteer to fight the fires.

Linda's reaction, she recalled many years later, was on the bratty side coming from a limited teenage outlook.

"The rest of us have to stay there at this house and to me this is, 'God, is he kidding me?'" she said. "I mean I'm stuck out here. There's no phone. There's no TV. I mean, could it be any more dull than this? I'm fifteen and I'm surrounded by my family and my little brother, who must have been ten. He was just not a buddy at the time."

Linda could not know that the following hours were going to be filled with more adventure than she wanted. A few minutes after her father and brother set off, they were back at the front door yelling, "There's a tidal wave coming! We have to get to the garage!"

Driving toward town, Doug had his window open when a friend yelled to warn them that a tidal wave was roaring in. Linda has always speculated that if the window hadn't been rolled down, they might not have gotten word about the tsunami, and the family might have perished. Her father didn't even turn the car around; instead, he threw it into reverse and raced backwards down the street.

"It was really just a matter of split seconds," Linda said. "There was no second-guessing."

The family evacuated in what seemed like a minute flat. Linda made sure she grabbed her bag with all of her prom stuff as it had not yet occurred to her that there might not be a prom the next day.

Everyone scrambled out the door, ran around the back of the house, and climbed up onto some oil barrels to reach the roof of the detached, flat-roofed garage. The garage was not a sturdy one, being open on the sides, almost more of a shed. A brand-new Cadillac was parked there, and to this day Linda vividly remembers the fins sticking out. From their vantage point, they could see Resurrection Bay and the approaching wave.

"We could tell that we were in a lot of trouble," she said.

The garage certainly would not hold up if a tidal wave rushed into town. One by one they boosted themselves up from the garage roof to the top of the ranch-style house, perhaps two feet higher and separated from the garage by a foot or two. Desperately they worked together to bridge the gap and cross over to the roof, one by one. In her arms Linda held the baby, Doug Jr., who was less than a month old.

She could see the big wave approaching, "dark and ugly."

The family made it to the roof top and lay down, holding onto the pitch of the roof. Two seconds later, the wave arrived with tons of force, slamming into the house and lifting it off its foundation.

"We were just like a little cork," Linda recalled. "It was like Dorothy in *The Wizard of Oz*. I've never found words to describe what it was like. We were spinning really fast."

Linda later described the perilous moment in an interview with the Associated Press.

"It took the garage and tore it all to pieces," she said, "and tore the porch and all the bedrooms off the house."

With seven people clinging for their lives to the rooftop, the structure was suddenly floating, the only home in the subdivision that had not been destroyed. It caught the wave like a surfer and floated inland, intact, toward a bridge connecting Seward with its airport.

"We were real high on the water," Linda said. "We weren't pulled back out into Resurrection Bay. We were pushed into all of the trees near the airport. It was amazing. The trees, with their huge branches, really could have speared anybody or swept anybody off, but they didn't."

After floating with the current for about fifteen minutes, the house, or whatever remained of it, lodged in a copse of cottonwood trees, ending the crazy, unbelievable trip.

"We took a wild, wild ride," Doug McRae said years later when interviewed for a documentary about tsunamis. McRae used some television wire to lash the remnants of the house to the trees. As the family endured on the roof, the water roiled past, carrying pieces of other houses, loose mattresses, even dogs.

"Everything floated by," Linda said. "My dad's car floated by. We never drove it again, but days later it was found out in all of the rubble, and our little black cocker spaniel was in there alive."

The McRaes were not going anywhere. More waves washed in, creating more terror and doubt. At one point, Linda's older brother begged his wife, Joanne, to climb higher in a tree and to take the baby with her.

"She wouldn't do it," Linda said.

More waves brought a frothing ocean to their shelter. About 11:30 p.m. that night, about six hours after the first jolt, another large wave rushed in and caught them by surprise, rekindling fear, as the water lapped at their feet.

"They didn't come like the big one (first one)," Linda said. "You'd hear like a high whistle, like wind. It would rush in and it would creep up and you never knew when it would go back down. This house is just being barraged every time this happened. It was dark and started snowing, and the water crossed my ankles. It was getting pretty darned high. Nobody was hysterical, but people were very tired and the night was endless and I think there was a lot of deal-making with God."

In between waves during the long night, each one in the family had their own private thoughts. Linda's thoughts were a lot more sober than they had been earlier in the day, when all she could think about was the prom.

"I realized that if we had died in the initial wave, it would have been so fast and so quick we wouldn't have been found," she said. "We would just be missing people. Nobody would have survived."

The family talked among themselves for reassurance, too. Scott McRae recalled how the family had come to Alaska from Montana during World War II. He and his wife were young newlyweds. Having no family in Alaska, Scott always kept a significant amount of cash in a money clip so he could buy a plane ticket fast in case they had to fly to Montana for a family emergency.

Linda remembers him saying, "All this money, and I can't save my family." He still had a money clip in his pocket.

Once the waves stopped battering them, someone climbed down into the house and returned with a can of peaches for the baby, who was protected from the cold air by a quilt and warmed by two candles. They tore insulation off the inside of the roof and wrapped themselves in it for warmth.

Linda kept saying in her head that rescuers would be out searching for them, as if they were the only people in dire straits.

During periods of calm, the family could see the flames still shooting into the air at the port and occasionally hear or see planes flying overhead.

Saturday morning finally arrived and, with it came the realization that they were in a precarious place—holding onto a roof for survival on a house that could either break apart or float away at any time. The difficult and disputed decision was made to seek help. The situation was exacerbated by the fact that when Joanne or Doug had hurriedly grabbed their diaper bag for little Doug, they had taken the wrong bag and had no baby formula.

Doug volunteered to get help, but the rest of the family argued against him going alone.

"Nobody wanted him to do it because it was scary, but he decided to leave," Linda said. "We just didn't know what was going to happen and what he was going to find."

Still, Doug slithered off the roof and made his way into the remains of Seward, walking about three miles. Hours passed and no one on the roof knew what was going on. When Doug reappeared, it was from the opposite direction, and he was accompanied by representatives of the American Red Cross, who had come part way in a jeep. After twelve hours on the house-turned-boat, the family was led away on foot, walking through deep mud where the water had receded.

"It was real mucky, deep, deep awful stuff," Linda said. "We had to walk toward the bay, and I remember not really wanting to do that."

When they reached temporary shelters in Seward, they read the names on a list of missing persons.

"And we were on that list, all of us," Linda said.

The family stayed at the local armory for a day while everyone was checked over; then the McRaes were allowed to go home.

Of course there was no junior prom that night. School was called off in Seward for about a month. After classes resumed, a substitute prom was scheduled, and Linda attended with her date. She was not wearing the dress she had so tenaciously clung to until the lifesaving leap to the roof. It was lost in the tidal wave.

"My dress was never found," Linda said. "It's out there haunting me. I bought a new dress. I don't even remember the color."

Two years later, she married Steve MacSwain.

Over time, as remarkable and unique as the real McRae story was, embellishments have taken root with the retelling by others. Distant friends or acquaintances told fragments of the story that never happened. One version had the house being sucked out to Resurrection Bay and then hurled back on land. That did not occur. Another version had Linda wearing her prom dress—not true.

The intense experience changed her, Linda said.

"I really always felt lucky to be from Seward, and obviously to survive that experience. I think it made me very, very strong. My entire life I've been strong." she said. "These reality TV shows just kill me. Come on! You're gonna make me eat a bug! Are you kidding me? It's just so stupid."

The McRaes' reality show was quite enough for one lifetime.

9
Valdez

When the 441-foot cargo ship sailed into port, the children ran to the dock. The crew made a habit of tossing fresh fruit or cookies and baked goods to the children of old Valdez. By grim happenstance, the SS *Chena* had been docked for only a short time when all hell broke loose.

It was here, in Governor Bill Egan's hometown, where the quake and tidal wave did their greatest share of heartbreaking damage. The epicenter of the most powerful earthquake in North American history was located fifty-six miles west of Valdez on Prince William Sound at a depth of twenty miles.

Valdez, with a 1964 population of about 1,200, was referred to as "the Switzerland of Alaska" because it was surrounded by the picturesque Chugach Mountains and the water of the Port of Valdez leading into Prince William Sound. It was one of the snowiest communities in the world with an average snowfall of more than 325 inches.

In this small town far from Alaska's population centers, the devastation was overwhelming. It wasn't until it experienced the tumult and destruction caused by the earthquake that residents realized that in 1898 the town had been built in such a precarious place.

"It was quite a big event when the supply ship came in," said Fred Christoffersen, a Valdez resident who was twelve in March 1964.

On the most grimly historic day in old Valdez's history, Christoffersen was one of the many children who dashed to the dock to greet the fourteen-thousand-ton *Chena*. That is where he and some friends were when the rumbling started.

"We'd go down and the cooks would always pitch out apples and oranges to people standing on the dock," he said. "Everybody would go down there and collect a bunch of apples and oranges and usually take off right away because the dock was such a busy place with all of the forklifts and longshoremen running around working."

Christoffersen had gotten his fruit and was stepping off a wooden section of dock onto solid land, a gravel road, when the terror began.

"I didn't actually know what was going on," Christoffersen said. "It sounded like a loud explosion to me, and I turned around and looked over the water to see what blew up. It sounded like a bomb went off. And then the ground started shaking and the young fellow with me had come from California and knew what earthquakes were. He hollered, 'Earthquake! Run!'"

The road was icy, with some snow on it, but they were able to keep their feet as they started running away from the dock. When Christoffersen turned around to look back, he was stunned to see the *Chena* bobbing up and down with the waves and buildings beginning to crumble. The tops of power poles were snapping off. The frozen ground began ripping apart nearby.

It was a sensory assault. Everywhere Christoffersen looked, something was falling apart as the earth shook. The noise was tremendous. The ground was tearing apart. Things were breaking. For more than five minutes as the earthquake roared, it was hard for him to pick out individual sounds.

Christoffersen's friend kept running and yelling, "Run! Don't look! Run!" He got about half a block ahead of Christoffersen, who was mesmerized by the astonishing sights and sounds.

"I kept turning around watching," he said. "I got up by the building that they called the village morgue. It was actually just a bar and restaurant, I guess.

Then the whole road tilted to the south and a semi-tractor on the other side of the road slid across in front of me so I had to stop and wait for it to go by before continuing to run inland on the high side of the road. That village morgue started falling apart wall by wall until it crumbled down and I got to the end of the gravel road where it met the land mass. There was kind of a jetty there. Then a big crevasse opened up. I slowed down to stop, but I figured, 'Boy, if I stop now, I'm never going to make it,' so I kept going, picked up a little speed and jumped that crevasse. I don't know how wide it was, but it seemed like six feet."

Behind Christoffersen, the earth was moving in waves. It unleashed pressure ridges of frozen dirt and ice. He had to keep going to outrun what seemed like the earth chasing him. "I knew that if I could stay ahead of that stuff, I'd probably be okay," he said.

Christoffersen was running to save his life. He wasn't headed home or to a specific destination; he was just running. "I'm still down on the waterfront," he said. "I haven't even got any idea if I'm going to make it home."

For good reason: Christoffersen witnessed the horrifying scene as the *Chena* ripped from its moorings and the dock was swamped, washing away more than thirty people.

"It actually went to pieces when a big wave hit the ship and pushed the ship onto the dock," Christoffersen said. "It exploded it. As a young kid I had never seen anything like this before and it was terrifying. You know at the time a number of people have died, but it didn't really register how many people lost their lives."

Christoffersen remembered seeing a happy Jim Growden on the dock. Growden, twenty-eight, had one of his little boys, either David, four, or James Jr., two, riding on his shoulders and the other holding his hand. All three were killed. Growden, who was from Fairbanks, was going to move back at the end of the year to coach the Monroe High School basketball team. The baseball field that is home of the Alaska Goldpanners is named after him.

Christoffersen and his friend kept running away from the water. The other boy lived on a corner near the museum. A man named Ed Cronke, whom Christoffersen knew, pulled up in his panel truck and yelled for him to get in. At first Christoffersen refused because he wanted to get home and see if his parents were all right. Cronke replied, "Don't worry. They're already gone."

Cronke drove him to an evacuation way station six miles out of town. Christoffersen's family had come and gone from there, too, moving on to safer, higher ground at the Thompson Pass state highway department maintenance station twenty-seven miles from Valdez. If it had been much earlier in the year, with winter still in full grip, it would have been a challenging drive. Snow is measured in feet in the 2,805-foot Thompson Pass. Christoffersen's parents and five sisters had been transported farther north on the Richardson Highway to an aunt's house in Glennallen, 115 miles away.

Young Christoffersen's family had been told that he last was seen on the dock. They believed he was dead. Everyone knew about the loss of life on the dock.

"They had already put out that I was one of them that had died," Christoffersen said. "When I walked into my aunt's house in Glennallen, my mom, Thelma, turned white as a sheet and wanted to know if that was really me. She had it in her mind that I was gone."

Everyone was glad to be reunited, but being away from their home and hearing alarming reports of damage and depressing stories about the deaths of friends and acquaintances, Christoffersen said they didn't talk about what he had seen. Later, he didn't feel like talking about it—not when he got back together with his family and not during the year that the family spent in Glennallen, either. It was only much later, when he was in his fifties, he said, that he talked about the day the dock was swept away.

He still becomes emotional recalling the disaster. As he became older and more reflective, Christoffersen said the scene implanted in his mind bothered him more than it did when he was a twelve-year-old living the experience. Christoffersen's father was a fisherman who went out to sea often; his son fished with him every summer. After the quake, without ever voicing his thoughts, he always worried about the ocean and what it might do to them.

"I didn't want to go out after seeing what had happened," Christoffersen said. "After seeing what happens when you have an earthquake and how frequent earthquakes are in Alaska, the older you get and the more you read and learn about this stuff … you know an earthquake of that magnitude should be a once-in-a-lifetime deal. But I never went back to that old town site where that dock was for maybe thirty years."

Christoffersen's family never talked about moving away after Valdez was assaulted by the quake and tidal wave. As an adult, Christoffersen, a heavy-equipment operator and truck driver, never thought about moving away to another community, either.

"I'm one of them people that don't move, I guess," he said. "Some people stay in one corner of the world for a hundred years. I guess I'm one of them."

THELMA BARNUM, who divorced Fred Christoffersen's father and remarried, had nine children at the time of the quake. Most of them were at home with her in Valdez as she wound up her old wringer washing machine. She was draining the washtub to do laundry and getting supper ready. Fred was at the dock with friends. A daughter, Christine, five, was playing with a girlfriend down the block.

The earthquake sounded like thunder at first, and then the ground started shaking. Right away Thelma was certain the Russians had dropped a bomb.

"That was in the cold war, and I think quite a few people did," she said, years later when she was seventy-nine.

Thelma was thrown to the floor and could not get up. The first thing she did was crawl to the washing machine and unplug it because the wringer kept twirling around. One daughter was ill with the mumps. The other children were dressed to go out and play right before dinner, so they had their winter clothes on, saving time as they prepared to evacuate.

Then-husband Ken Christoffersen was holding the baby, Kenny, and was mesmerized by tall birch trees and telephone poles bending back and forth. Thelma grabbed the tail end of Ken's shirt to get him moving. The Christoffersens' car was not running and they weren't sure how they could escape the neighborhood.

"It was a terrible time," Thelma said. "I ran along a snow bank carrying one of the small children. It was crusty, and I had bad legs. I didn't know what to do. I said to my husband, 'Throw all of the kids on top of the roof. If it floods, we might be able to float.' We were starting to do that when all of a sudden the earthquake stopped and closed up the cracks. At one point, one of those cracks was coming right towards me. My husband was in the house getting something and we were getting ready to go, although I don't know where we were going to go. I was standing out there and this crack came right at me, with water spurting up. I couldn't move. I had two little ones beside me and the baby in my arms. I said a prayer and all of a sudden it stopped six inches, maybe a foot from me, swirled around the house, took its path and went through the brand-new house behind us that had a concrete foundation. The house cracked in half. The old wooden house we were renting stayed intact."

Soon after, friends drove by and whisked the family away to safety.

"But I didn't want to go," Thelma said. "I had a wrestling match with my husband because I wanted to find my other children. I was frightened, but I wanted to find my children. I couldn't find the little girl who had been out playing. My husband shoved me in the car, made me stay there, slammed the door and told the driver to go. I was angry for about two days."

She demanded that they find their little girl, who as it turned out was at a neighbor's house with playmates.

"I probably would have gone to the right house," Thelma said. "But at the same time the earth was breaking open, and these big, twenty-five- to thirty-foot spurts of water were spouting out of the earth. Water was flooding the streets. I was going to try and go around to the places our daughter might be, but my husband wouldn't let me out of the car."

When Ken got most of his family safely set up at Thompson Pass, he hitched a ride with a truck driver he knew back to town to look for the other two children. At about 5 a.m. he returned with his missing daughter, but not his missing son.

"He was a fisherman by trade and used to the ocean, but I think it scared him," Thelma said. "He was safe, but he said it was only by about three inches. There had been three waves in succession."

They spent one night at the camp before moving on to Glennallen.

"People told me, 'I saw your son,'" she said. "Pretty soon I began to wonder did they really see him or see somebody else. Everybody was in turmoil. I didn't really know. My mother said that he was gone, and I said, 'No, mom, I'm not going to listen to you. My son is alive. I know he's alive.' I just kept wishing and hoping."

After two days in Glennallen, Fred still had not been seen. Most of the family was gathered in a trailer behind the lodge when a few of the younger kids burst into the room shouting, "Fred's here, mom! Fred's here!"

"My knees almost buckled," Thelma said, "but I ran out to see him, and of all things, he had our bird, our bird cage, the family parakeet. This poor little kid, no parents at home, and he's rescuing the bird. It was very emotional."

Thelma Barnum said the entire experience was emotional and stayed emotional for a long time. Everyone who survived the earthquake and the tidal wave in Valdez had his or her own personal experience and a story to tell.

TOM MCALISTER, who later became president of the board of directors of the Valdez Museum erected to commemorate the earthquake and those who perished in it, was one of those people. After years of thought, he distilled the experience of living through the catastrophe into two words: "Pure terror."

McAlister was working construction in 1964. At 5:36 p.m. he had arrived home from work, ready for dinner. McAlister and his wife, Gloria, who had a six-week-old baby boy named David, had developed a ritual of having a special dinner on Friday night once a month. They called it "steak night," and March 27 was one of those Fridays.

The raw steak was set out, ready to be cooked, and was still sitting in the same place three days later after the McAlisters' desperate flight to safety.

"We had earthquakes all the time of various sizes, but nothing that lasted very long or was real strong," McAlister said. "One winter I lived in a house with a state trooper and one night we had an earthquake. He said, 'We better go out and check the town.' We got dressed about three o'clock in the morning, went out and looked at fuel tanks and power lines and everything else. There was no damage, but that was one of the things you faced."

This earthquake was different.

"It started and my wife looked up," McAlister said. "I said, 'It's just an earthquake.' It started again and it didn't quit. It was, 'This is very unusual.' I was sitting down and I got up and looked out the window. My wife started to panic; I picked up the baby and I grabbed her and we stood under a door frame. That was something I learned about in school. I grew up in western Washington in the fifties and they had an earthquake that shook the tallest building. So at my school they had an earthquake class and that was one of the things they told us, to get under the strongest point of the house, which is usually the door frame. So that's what we did."

When an earthquake of this historic magnitude comes along, earthquake education becomes theoretical. The McAlisters lived in what he thought was a sturdy house. It was built on posts with a solid basement, but the shaking moved it about two feet off the foundation. As a young married couple, the McAlisters didn't have many possessions in 1964, he said, so they didn't have much to be damaged or destroyed.

The McAlisters lived a short distance from Gloria's parents, but in the upheaval, the 3,500-gallon town water tank toppled and spilled water into the streets, and pipes fractured. Tom's father-in-law and brother-in-law were in Anchorage, so Gloria's mother was alone. Tom McAlister's basement filled with water and his heat went out, so initially he and Gloria and David took refuge in his in-laws' home. Then came word of the threat of an approaching tidal wave, and they evacuated to the six-mile camp outside Valdez.

"We sat on a hill for two or three hours and then came back to town," he said. "There was virtually no communication. Our only communication with the outside world, after a day or two, was our lone ham radio operator."

Like Seward, Valdez was initially hit by a wave caused by a nearby landslide. The massive wave washed away the dock after an estimated ninety-eight million cubic yards of earth slid into the harbor.

"We were concerned with our own needs and other peoples' needs near us at that point," he said. "About eleven o'clock that night there was an extreme high tide that came in with the backwash of the tsunami that went south. That's when water got into an electric power meter and set some buildings on fire. It actually set oil floating on the water."

The Standard Oil warehouse, Club Bar, and other buildings went up in flames. At that point Valdez was evacuated again. Everyone moved to the twenty-seven-mile camp, where about three hundred people were housed the first night after the earthquake and served hot soup and drinks around the clock. At 6 a.m. the next morning, McAlister returned to town to pick up clothes for everyone in the family. Walking along the beach, he saw the destruction of buildings and the waterfront. It was a sobering sight.

Two days later, the McAlisters returned home and began to rebuild their house as so many others sought to do.

WALTER AND GLORIA DAY published the *Valdez News* that sustained the small community by feeding it the basic news. Thinking back, Gloria can still see the moving piano. She had been up very late Thursday night, in newspaper vernacular, putting the weekly newspaper to bed. By late afternoon Friday, she was having trouble staying awake. So she took a shower, changed into her pajamas, and was about to take a nap. Then the house started shaking, and Day was startled to see her piano rolling back and forth across the living room.

"I dodged around that and just about that time my daughter, Wanda, who was about fourteen, and a friend of hers walked into the house," Gloria said. "They had just come up from the dock. That cook on the *Chena* always made goodies for the kids and he passed out apples, oranges, and cookies. My husband Walter came in."

The front porch of the three-story home had a high view. When the house started shaking, Walter and three of the children (two older children were away) went outside to the porch where from a distance they could see the *Chena.*

"It was up in the air and we could see the stern of it and the rudder and everything," Gloria said.

Later reports indicated a wave had lifted the ship thirty feet.

"That made us conscious of how terrific this thing was," Gloria said. "When you looked down the street where there was a two-story bar, a log building, the ground was rolling so much that the bar dipped out of sight. Then we noticed the streets were getting water in them."

Everyone knew that Valdez had been built on a water table that was only a couple of feet below the surface of the ground. Most houses didn't have basements, and the Days' home had a basement that was basically above ground, where the furnace was. As publishers of the *Valdez News*, the Days also had their printing press and other production equipment on that level. The business was in their home.

The Days learned that Walter's sisters' husbands, Bob Harrison and Doug Granger, had been killed on the dock. The Days owned a homestead cabin about four miles from town, so they jumped into their car and were about to flee to it when their sixteen-year-old son Pat drove up. He followed them.

"I was absolutely scared stiff," Pat recalled.

The road out of town was in bad shape. Water was everywhere. Pat remembers geysers spurting out of the flats, some hurling water fifty feet into the air. As he turned around to return home, he stopped to inspect a concrete bridge crossing runoff from Valdez Glacier. There was a gap on one side where it had pulled away from the road bed about six feet wide. He said he just backed up the car, gunned the engine, and raced over the gap in the air, as if jumping it.

The family made it out to the cabin. It was a stunned and sad group consisting of Gloria and Walter; three kids, including the youngest, Linda, twelve; the two women who had lost their husbands; and Pat's girlfriend, whose father had been at the dock, too, and was thought to have died.

"We were still looking for information," Gloria said. "I'm not sure how we found out, but we knew that the dock was gone and that the two brothers-in-law were gone. All we had was a one-room cabin, but we stayed there that night."

They built a fire, made coffee, and ate sandwiches they brought from town. The mood was somber.

"You can't believe how really sad it was because here's three people that knew part of their family was gone, their husbands and dad," Gloria said.

After a night at the cabin, when it became apparent that Valdez was in ruins, the Days evacuated to an elementary school in Copper Center, where they stayed for about a week. Then they rented a cabin in Glennallen. Later, Gloria Day was appointed postmaster of Valdez, and the family moved back to the community. The couple's newspaper days were over. They never published an earthquake edition. The competing *Valdez Breeze* was shut down until the following August. However, in mid-April, a group of citizens began producing a slender mimeographed newspaper every few days called the *Valdez Earthquake Bugle.*

The Days' house had shifted on its foundation, and when they returned to Valdez they moved into a trailer provided by the state.

Everyone in the town was touched by tragedy, and although people tried to adjust and adapt, there were emotional scars.

"It was very draining," Gloria said. "It was a shock. A young man was supposed to get married on Sunday and he died on the dock. The basketball coach (Jim Growden) where our kids were going to high school, he and his two boys were down there and they were gone. They're your friends and all of a sudden, they're all gone. I talked to a doctor a few months afterwards and he said even though you carried on there wasn't a person in town that wasn't in a state of shock."

THE DEATH TOLL in the Valdez area was thirty-one, all but three of them killed on the dock. The dead included two members of the *Chena* crew and a man in his boat trying to reach a cabin on the bay.

The dock and waterfront vanished so swiftly that at first the loss was hard to comprehend. The arrival of the *Chena* was ordinarily a happy occasion, not only for the kids getting their treats, but because the ship brought freight to the isolated community. Orders placed long before were being filled, including orders for Easter.

The Valdez Dock Company owned the city dock. The area included warehouses, a cannery, a packing plant, a bar, a slip for the ferry that ran between Valdez and Cordova, and a small-boat harbor for the local fishing fleet. The heart of the tiny business district was two blocks from the water.

John Kelsey, one of four owners of the dock company, said everyone knew that Friday was going to be "an action day" because the *Chena* was due in about 5 p.m. Once the ship arrived, it was going to be a hectic twenty-four hours of unloading to warehouses until the ship sailed off.

When the ship docked, Kelsey said he made sure the captain got paid, and Kelsey was invited to eat dinner onboard. Kelsey declined because his wife was preparing a dinner of "Valdez large shrimp," and his wife's parents would be there, too. He and his family, including a seven-year-old daughter, were sitting down to eat when the first tremors were felt.

The shaking, Kelsey said, "kept getting harder and harder and harder, the noise ominous."

Kelsey said his mother-in-law tried to control a bouncing coffee pot but was thrown to the floor. His father-in-law, stumbling into a hallway, kept repeating,

"Oh, my God, oh my God, this is the end, this is the end." Kelsey said the only time he ever heard his wife speak rudely to her father occurred right then when she told him to shut up.

When Kelsey made it outdoors, he looked toward the waterfront. The docks were gone.

"I could not believe it," he said. "I was in disbelief. To this day I can remember thinking 'This cannot happen.'"

Kelsey's home tilted off its foundation, and the family evacuated to Fairbanks.

INDIVIDUALS EXPERIENCED widely different forms of terror. John Bedingfield, who owned the Valdez Hotel, stumbled into one of the cracks in the ground, much like a crevasse on a glacier, but a powerful spurt of water saved his life.

"The force of the water was so severe, so strong," he said, "that it threw me right back up to the surface."

At Harbor View Nursing Home, patients were sitting down to dinner when the building shook. People were thrown to the floor, unable to get up. Ann Kelsey, who worked at the hospital, reported that the concrete in one wall opened and closed.

"I thought it was the last day on Earth," she said.

The bell at the top of the Church of Saint Francis Xavier kept ringing loudly, unattended, as the earth shook.

Water rushed into town as far as McKinley Street, a few blocks from shore, and the street was cluttered with debris and chunks of ice after the water receded.

On the outskirts of town, Basil Ferrier and his son Delbert were working a hand-logging operation on the shore of the Valdez narrows. They traveled there in their thirty-four-foot boat, the *Falcon,* which they had difficulty anchoring that day. They tied the boat to a tree with a two-hundred-foot cable, which proved to be their salvation.

When the quake hit, a wave rolled out from the shore where they stood, carrying their skiff with it. However, on the next wave the boat returned. They jumped into the skiff for a very rough ride to the *Falcon.* As they reached the larger boat, the skiff was quickly filling with water and they were in waist-deep water as they made a leap to the *Falcon.* The skiff sank.

The Ferriers got the boat engine started and ran as fast as they could. A wave rolling to shore appeared to be about fifty feet high.

"I was busy with the controls," said Basil Ferrier, who was urgently urged onward by Delbert. "I felt that if we could beat it out of the narrows and into a bigger body of water before it hit us, we might have a chance as it would flatten out a little. My son's next report was that the wave had taken out the big narrows light, which is about thirty feet high and built with steel and cement. I felt then that all we had to do was wait, that we didn't stand a chance."

The Ferriers reached a wider area of water as the wave caught up to them. They had run just far enough and fast enough to escape the brunt of the wall of water.

"We had a rough ride," Basil Ferrier said.

Then he turned on the boat's radio and heard that Valdez had been clobbered. That's where their family was. It was dark by then, and the men had to maneuver the boat through floating debris, much of it apparently from the docks. Ferrier dropped the anchor line around a log, and they waded to shore, walking over the tops of automobiles. The family was unhurt.

"Our survival was nothing short of miraculous," Basil Ferrier said.

CONVERTED FROM a Liberty ship, the *Chena* nearly sank in the initial tidal wave. Captain M. D. Stewart of the Alaska Steamship Company had been looking forward to a hearty dinner aboard as the unloading continued. Suddenly he found himself trying to save his vessel. The ship was lying over to its port side, in danger of going down.

As the dock disappeared, the ship came down from its violent tilt in the same spot where only minutes before wood and planking and people had been. The ship struck bottom and bounced up. The waves helped right the ship each time. Fishing boats and yachts that ripped loose from moorings crashed into the *Chena.* Inside the vessel, two stevedores were killed by heaving cargo and equipment.

Stewart called for power and desperately sought to turn the ship to sea. One wave turned the bow away from land and that helped prevent the ship from grounding and aided its propulsion into open water.

"It is a miracle that the vessel survived," Stewart said later.

It was a day when more than one survivor who had been up close with the waves spoke reverently of miracles.

When things calmed, Stewart anchored about a half-mile offshore and sent out an SOS for medical help. One of his men, Jack King, had suffered serious chest and leg injuries from being tossed about. Dr. Clarence Davis, chief physician at Harbor View Nursing Home and Valdez's general practitioner, and nurse Betty Melynn took a motorboat to the *Chena* and worked to save King's life.

The *Chena,* surrounded in oil-slicked waters by debris from the dock and from a town that had virtually washed out to sea, departed the next afternoon for Cordova. There, King was treated at a hospital. Both his feet were amputated, but he lived.

In interviews later, Second Engineer Carson Dorney and others recalled the disaster. He had watched the dock collapse and Valdez buildings begin to crumble, and he saw some go down with the dock.

"It was horrible," he said.

"Three men were spinning around in a huge pool," Dorney said. "Two went under right away. The other was floating on a piece of roofing. It swallowed him, too. The ship washed on sideways into where the dock had been."

"The ship was going up in the air," said Chief Engineer Chet Leighton. "Boom! Boom! It was like a rough elevator. I could look over the side and there was no water. We were way up in the air. People in town said the ship went so high they could see underneath it. All this happened in slow motion. People, buildings, everything went down in the hole (a crack in the earth). The hole was full of pilings and rubble. A wave came in and took everything else."

"I saw a lot of misery," said crewman Kenneth Wiper. "I've been all over the world, but I've never seen anything as terrible as that—people losing their lives in front of you."

As residents shuttled back and forth between the six-mile camp, Thompson Pass, and town, a core group of citizens sought to assess damage and restore some of the lost utility functions. Owen Meals operated the power plant and telephone headquarters where two generators still supplied power.

Mayor Bruce Woodford, John Bedingfield, Don Williams, a city council representative, and businessman George Gilson formed a supervisory team.

Although most citizens didn't know it, some long-distance telephone service was running, and Meals had made contact with Cordova and Juneau.

What no one had any control over were the fires at the Union Oil Tank Farm that consisted of ten-thousand-gallon barrel tanks left over from World War II. The fire burned for two weeks. It seemed as if the entire town was engulfed with oil spillage.

"During the earthquake an oil stove fell over and caught fire to the building, and that spread to the tank arm and it fed upon itself," said Kelsey, who at that time was serving his turn as chief of the volunteer fire department. "One tank would catch fire and start burning and then that would go to another tank, and it went from one to another, and we just gave up trying to control or do anything about it. I equate what I saw to what I saw happen during World War II (in the Pacific)."

Another fire, kept under much better control, ignited at the Chevron and Standard Oil tanks.

A temporary kitchen to feed those with no access to supplies was set up at the Switzerland Inn. Gilson, who owned a grocery store, brought in all the food he could salvage. Cooking was done on a Coleman stove, and water was brought in from the local power plant. Coffee, bacon, and eggs were served Saturday morning.

Later that day, a community kitchen was set up in the Harbor View complex, which had largely survived the disaster, and it continued operating for a month as people struggled to regain normalcy. Gilson donated large amounts of food from his undamaged grocery warehouse.

Word spread quickly of a need for shelters. Those driving beyond Thompson Pass were greeted by signs "Can take three" or "Room for two" on homes along the Richardson Highway.

Helicopters dispatched from Fort Wainwright near Fairbanks, 364 miles to the north, evacuated many women and children. Among those who had relocated to Glennallen, evacuated to Fairbanks, or went elsewhere, it was estimated that only about one hundred Valdez residents remained in the community twenty-four hours after the earthquake.

The Army moved in overland, as well, with 107 soldiers and officers arriving within a day, carrying fresh water, C rations, lighting, and communications equipment. A doctor accompanying the troops issued typhoid shots to the residents still in town. Entire streets in the business district were impassable, covered with smashed-up telephone poles, empty fifty-five-gallon drums, tree limbs, ice chunks, and other debris.

Two days after the earthquake, the mimeographed paper, the *Valdez Earthquake Bugle,* included the compelling story of the quake from a previously unheard source: Captain Stewart. Stewart said the *Chena* arrived at 4:12 p.m. and it was just past 5:30, when its stores were being unloaded, that those aboard the vessel felt the first stirrings of the quake.

"There were very heavy shocks of about half a minute," Stewart said. "Mounds of water were hitting us from all directions. I was in the dining room. I made it to the bridge (three decks up) by climbing a vertical ladder. God knows how I got there.

"The Valdez piers started to collapse right away. There was a tremendous noise. I had been in earthquakes before, but I knew right away that this was the worst one yet. I saw people running with no place to go. It was ghastly. They were engulfed by buildings, water, mud and everything. The *Chena* dropped where people had been. That is what has kept me awake for days. There was no sight of them."

In its April 10 issue, the *Bugle* thanked the Army: "Some of the first and most deserving praise should go to the U.S. Army detachment which has furnished twenty-four-hour security as well as much other assistance too numerous to mention. Many thanks and heartfelt gratitude to Major Bowers and his group of men, even though they have already returned to Fort Wainwright. We're very grateful and proud of our U.S. Army."

The newspaper also warned its readers: "Do not drink any other water than that issued by the Military." Seventeen issues of the *Bugle* were published.

AFTER DENNIS EGAN woke up his napping father, he did not see him again for days. The governor adjourned from his Juneau office to Anchorage as soon as he could to set up a temporary emergency headquarters.

Egan also reached out to members of his family for reports from Valdez. Although communication was hit or miss, on his first try Governor Egan was able to reach his brother Truck, then bartending at the Pinzon Bar, which had sustained heavy damage.

Bill Egan was told his sister, Alice Horton, had experienced trouble negotiating through the damaged areas to get home, but some local boys helped her. On Easter morning, two days after the quake, Valdez residents who had stayed behind were working with the Army on clean-up.

"I was so moved with the spirit of 'Let's get on with things,'" Egan said. "They were as busy as bees on Easter Sunday morning."

Bill Egan's family home in Valdez had been damaged beyond repair. But he was too busy being governor of all the people and didn't have time to dwell on Valdez at the expense of other communities. So many people suffered such great losses, including thirty one who lost their lives in Valdez, that talking about loss of personal heirlooms seemed trivial.

"We lost everything," Dennis Egan said. "Photos, stuff—you know, memorabilia. It was a small house. We were close to the oil tanks. We were only about a block away from the Chevron tanks and the dock. We were real close to the water and that stuff all disappeared."

The governor did not speak publicly about the family's loss. Dennis said his father "didn't want to dwell on it and he felt for everybody else because everybody lost so much. He said, 'Why the hell are we going to talk about stuff? Everybody lost things.' And we were alive. We were here in Juneau."

After Egan was re-elected, the family built a house in Anchorage to replace the home in Valdez. Dennis Egan was about eleven when the family moved to Juneau and then on to Anchorage, but he said his earlier years spent in Valdez were among the best of his life. He had a carefree, fun childhood where he could roam around the small community and play outdoors with his friends year-round.

"I have great memories of Valdez," he said.

10
Kodiak

Inside the Donnelly & Acheson Building, Joe Floyd was sorting mail at the Kodiak Post Office on Good Friday.

The office was about to close, and only a few people were still hanging around the small lobby. Outside, the sun was shining, and despite the usual expectation of rain, the weekend was full of promise.

When the earthquake struck, the building shook violently. Everyone ran outside, where they saw dust billowing from a rock slide on Pillar Mountain, the 1,204-foot peak that rises behind the city. Soon dust hovered over the city.

"You know, I really don't think anybody around me understood what was happening at first," said Floyd, who was also a teacher and oversaw sports and recreational programs. "Then, as it continued, we made that mad rush to the outside. We could see this dust phenomenon, but we didn't see any kind of evidence of damage."

Located about 250 miles southwest of Anchorage, Kodiak Island is the second-largest island in the United States. The Kodiak crab fishery has been the richest in the world, and king crabs from surrounding waters are prized seafood on restaurant menus. The Kodiak brown bear, a larger version of the grizzly, is a feared and admired animal in the wild.

In the 1700s, Kodiak was the capital of Russian Alaska. The island was part of the purchase of the Alaska Territory from Russia in 1867. The 1960 census put the population at 2,628. Alaska Natives date back seven thousand years on the island.

Unlike other southcentral Alaska communities where trees and light poles whipped around, not even the telephone wires seemed to be vibrating, though the power went out. When the quake ended, everyone went back into the Post Office and locked up.

Floyd, then a father with two young children, Max and Virginia, made his way home through the community, not seeing much in the way of notable quake damage. He and his father, Joe Sr., climbed into their old car with a dog for company and began driving around. They ended up along the shore at Mission Beach, where they noticed that the water had receded from the shore.

"It looked like you could walk to Woody Island, which is a good mile from the shore of Mission Beach," Floyd said. "It was usually a semicircle of white sand, but at that time it was black sand. We looked around and we didn't see a whole lot."

It was a warning they didn't understand.

They returned home where Floyd's wife, Carolyn, who later served as mayor in 2011, was making supper. Floyd recalls the menu was sandwiches because there was no electricity. It was about 7 p.m. by then, almost ninety minutes after the earthquake. The Floyds thought things were back to normal.

"We didn't know anything," Floyd said. "All we had was Armed Forces radio that was relayed from the Kodiak Naval Station at the time out of Anchorage. So we didn't have any radio. It was gone. And of course the telephone was gone, too. So we sort of had to exist within our own environment."

As the family sat at the table, sometime after 7 p.m., it was startled by a huge. frightening roar.

"It sounded like we were almost side by side with a couple of freight trains passing one another," Floyd said.

A tsunami was swamping the city. The Floyds dashed out of the house and climbed into their new station wagon. It wouldn't start. Switching to their 1938 Plymouth, the family, including the dog, drove to high ground at the high school.

"This roar continued on," Floyd said, "and we didn't realize the extent of the wave. Over the next day or two so much water went through, we continued to have tremors, and waves that came at different times."

When the first tidal wave rolled onto shore, Floyd's daughter, Virginia, who was five, was out playing. She came into the house and told Carolyn, "Mom, there's water out there." Carolyn Floyd looked at her daughter tracking in wet footprints and replied, "Virginia, you're getting water all over the floor."

Carolyn thought Virginia had been playing in a puddle. Some puddle. When Mom went outside, she saw the rising tide.

People were fleeing from all around the city, evacuating to the highest ground they could find. Many retreated to Pillar Mountain. Hundreds gathered at the high school, as well, not knowing precisely what was happening, but stunned as tidal waves battered the community. Toting sleeping bags, the Floyds joined others sleeping on the gym floor, in the classrooms and cafeteria, and on the auditorium stage.

Principal Ivor Schott was one of the saviors of the community for opening the school and getting his employees organized to help care for townspeople.

The high school became home for the Floyds for a week.

It was the waves of water more than the earth shaking from the quake that did most of the damage to Kodiak. Waves rolled into the community for about six hours and drowned the waterfront. The seafood-processing canneries were wiped out, as were many boats and three-quarters of all local industry. Fifteen people were killed in Kodiak Island communities from tsunamis.

A fisherman who had been out in the harbor before the waves rushed in told his story to the Associated Press. Norm Holm and his son Oliver were fishing from a skiff, a boat that was functional on most days, but hardly big enough and strong enough to withstand a tidal wave.

Holm's first inkling that something was wrong occurred when he could no longer control the steering.

"I couldn't make it respond," Norm Holm said. "It kept bouncing, bouncing."

He immediately retreated to shore, where he saw many boats normally tied up at the dock breaking free of their moorings. Among them was his own seventy-two-foot boat, *The Neptune*, which Holm had to secure. The water level rose and dropped, and then Holm was shocked by a "terrible roar of water coming up the channel, the frightening noises of docks and buildings collapsing."

Jumping off their boat, father and son ran for high ground. They made it to the base of Pillar Mountain and climbed from sea level to 150 feet before resting.

Later analysis indicated that when water in the harbor receded, fishing boats were left resting on the bottom. When the tidal waves rolled back in, they yanked loose the boats and carried some of them three blocks inland before depositing them.

The "seismic waves," as city manager Ralph Jones called them, "inundated a large portion of the city and caused a part of it to wash out to sea." Electricity, public utilities, and telephone service were all knocked out of commission. The newspaper could not be printed. In the city offices, authorities issued numerous communication bulletins.

These included:

"Persons wishing to be evacuated and who have homes in the states can make arrangements by registering with Mrs. Doris Simon in Room #5 at the High School."

"A pass system has been established for movement into and within the general disaster area downtown in order to prevent looting and to permit cleanup crews to work unhindered."

"A curfew has been imposed on the disaster area and NO ONE is allowed in the disaster area from sundown to sunup."

"Meals are being served at the high school cafeteria during the emergency."

It would not be an exaggeration to say that given the widespread destruction and the casualties, Kodiak was like a war zone where a city had been bombed.

JAMES BOYD WAS in the Navy, stationed at the Kodiak Naval Base. Boyd had just turned twenty and was planning a birthday celebration on the town to start the weekend. He had finished showering and was sitting down reading a newspaper in the barracks when his hands began to shake. His first thought was that a nearby boat was revving its engine. Then a seaman yelled, "It's an earthquake!"

Like Boyd, many of the men were either naked or in their underwear, having been changing to go out on weekend leave. The force of the quake knocked many of them onto the floor. Rushing outside, Boyd couldn't believe how so many women had materialized. Women were rarely seen around the base.

"There must have been eight or ten screaming girls and women within a few yards of our barracks," he said. "Of course, liberty was cancelled and we were ordered to muster."

Once he was dressed and back on duty, Boyd was assigned to drive a squad of Marines into the city proper to help take control of an out-of-control situation. The drive was not easy. Even though the base was only a short distance away, it took an inordinate time.

"Rock slides had blocked many areas and we had to clear the road before proceeding," he said. "When we arrived I couldn't believe the destruction. The streets were littered with everything from rifles to cash. Looting was already taking place. The buildings that were on the waterfront were all displaced and in the middle of what used to be the streets."

As more waves had crashed into Kodiak, Boyd said his company worked fifteen to eighteen hours a day to remove debris and help wherever they could.

Monica Maach Tiller, whose father was a chief petty officer at the Naval Base, was nine at the time. She recalled the family just finishing supper and her dad settling into his easy chair to read the newspaper.

"I had just walked into the living room and stopped dead in my tracks," she said.

Soon after, when the shaking ceased and a warning of an approaching tidal wave was issued, the family moved quickly to higher ground and spent the night with strangers.

"We stayed the night with a family we didn't know," Tiller said, "the children sleeping while the parents stayed up all night gleaning news and waiting to see if we would have subsequent quakes or tidal waves."

KODIAK IS the major city on Kodiak Island, but there are several villages around the island dotting the shoreline. Those small communities were hit very hard by the tidal waves. Afognak, Old Harbor, Kaguyak, and Ouzinkie were devastated by the combination of earthquake and tsunami.

A woman named Betty wrote a letter to a friend in April that was preserved as part of the historical record. In it, she described the damage she saw in Kodiak. She was at home—a house located on high ground—when the earthquake hit "with a jolt," she reported. She was with a barefoot man named Jack, and each of them tried to stand in a doorway for protection while "canned goods (were) falling wildly to the floor."

Then the couple went outside for the duration of the quake.

"A chimney fell from the Taylor Apartments in front of us," Betty said. "I thought, 'I must tell the owner real soon because of fire danger.' An hour later he had nothing to hold up any chimney."

Back inside her home, Betty began cleaning up broken jars of mayonnaise and pickles. Initially, Betty and Jack had phone service, but soon enough their phone went dead, and they were alerted to the tidal-wave threat by someone coming through the neighborhood with a loudspeaker. Their view of the bay and the turbulent waters was scary.

"The water came up slowly the first time," Betty said. "The boat harbor dock (was) covered (and) then about twenty cars were covered. The next time I looked, the small house by the pier was covered."

The couple heard a cry for help, and Jack went down the hill as houses began to lose their grips on their foundations. He guided an elderly man named Frank back to the couple's house. Betty guessed that by then three or four waves had crashed into the waterfront, ripping boats away from their tie-ups.

"It hit the boat harbor and it was like a thousand guns going off as moorings and lines snapped like toothpicks," she said. "Everything moved toward town. The boats hit the stores and the stores hit each other and they all moved up the creek away from the bay."

Betty, Jack, and Frank watched the city break into pieces, including Frank's house. The water rose to the level of the couple's front yard but never got into the house. Inside, they had a fire going and slept in their clothes. Periodically, other neighbors who lost power showed up and warmed themselves by the fire. They all snacked on cheese and crackers together.

When the sun came up the next day, the sight of a wrecked Kodiak was demoralizing.

"A big barge sat beside the main street, partly on the sidewalk where (the) liquor store should have been," Betty wrote. "The street in front of us was packed solid with debris and houses all in the wrong places. The military was in charge downtown and we could not go down. The whole ocean was very high and stayed high for several days. Buildings floated all over it."

The crabber *Seleif* came to rest next to the old schoolhouse, still carrying a full load of three thousand live crabs that had to be thrown away. Jack worked eleven hours a day on the cleanup. Betty stayed home.

"No lights, no heat, no radio," she wrote to her friend. "We at our house had water right away. I stayed at home tending the fire, serving coffee and warmth to our cold neighbors, and listening to rumors. It was wonderful when the power came on—about seven days for us … we were really lucky."

There was a lot of discussion in the community of what would happen to Kodiak as each high tide in April brought more water into the area where buildings once stood.

"The talk right now is of filling the lowlands and building a sea wall," Betty said. "If they don't do that many businesses will have to move. We are still having mild quakes all of the time."

Kodiak lost eight lives and 158 homes in and around the city and the ground.

LATER, THE U.S. Geological Survey studied the earthquake and interviewed many survivors.

One unnamed Kodiak survivor told the USGS, "It went forever, never losing strength. My house must have moved a quarter of a mile from its original spot. I nearly went insane. Many around me threw up from nausea and violence. Babies went to the hospital for head trauma rather than cuts and scrapes. I was up on the hill, only to find a few chairs and sheets where my house used to be. I would never to my worst enemy wish this on someone."

Another local resident, Eddie Opheim Sr., his wife, and their four children survived the violent shakes of the earth unhurt and then jumped into a Jeep to head for high ground in advance of the tidal wave. They were above the danger zone, but in the night they could hear the tsunami doing its worst.

"We lost a sawmill, both boat shops, our barn and cattle, and our home," Opheim said. "We lived in an Army tent for a while until I built shacks out of scrap lumber that was left on the beaches."

Meanwhile, the Floyds were displaced from their home for a while, too. They remained at the school for a week and then moved in with friends in Baranof Park for a few days. About ten days after the earthquake, the Floyds moved back home to 1827 Mission Road, a house that Joe Floyd still owns. The house abutted Potato Patch Lake, which it shared with the Beach Comber, a popular nightclub.

When Floyd returned home, he took a walk out back and was amazed by what he saw.

"The lake was filled with all the boats and all of the trailers," he said. "The Beach Comber was demolished. There was debris all over that lake."

It was a make-it-or-break-it time for many people. Floyd had graduated from the University of Mississippi and went into the Air Force. He married his wife in Portland, Oregon, and took a job teaching school in Kodiak in 1955 with the game plan being to stay for two years. By 1964, however, Kodiak was definitely their permanent home, and they were still there nearly fifty years later. Carolyn Floyd said they never gave serious consideration to leaving.

School resumed in Kodiak in a few weeks, and Floyd went back to work teaching. The staff did everything, sharing all kinds of tasks, because there was so much to be done as part of the cleanup.

"We had to start the cafeteria lines and serve regular meals and we took shifts dish washing and everyone seemed to have responsibilities," he said.

The determination to rebuild overrode everything else at the time, Floyd said, with people helping people and the Salvation Army doing yeoman's work.

"They were everywhere," Floyd recalled. "We had the National Guard mobilized, the Marine Corps, Navy shore patrol, which came from the base, but I never saw any looting or vandalism. The Salvation Army became super-sized as far as the community of Kodiak was concerned."

Weeks later, there were still aftershocks, and any time one was felt, classes at school were edgy. One aftershock was big enough to startle Carolyn Floyd's high school class into running out the door.

"Boy, the kids took off out of the room and went out of the school," she said. "When they came back in after, I said, 'Well, boys, I'm glad that you let the women and children go first.'"

They had done no such thing. What they had learned from the Good Friday Earthquake was that when a big quake hits, head for the hills.

11

Far-Flung Ruins

The Natives of Chenega Bay had lived on Prince William Sound for thousands of years. The small village of twenty or so houses about one hundred miles east of Valdez was a stronghold of Russian Orthodox believers, who worshipped at a church that was one of the few public buildings in town.

The residents were Eskimos, Athabascan Indians, and Aleuts who, as well as being commercial fishermen, hunted seals, sea lions, and deer in a culture rooted in subsistence hunting, even if there was a grocery store in the village. Many communities disrupted by the Good Friday Earthquake and subsequent tidal waves lost electricity and running water. Chenega Bay had neither before the disaster struck.

When their houses shook with the onslaught of the earthquake, people recognized what it was. What they didn't recognize right away was the impact on the coastal waters. The water in the bay began to recede, a remarkable sight that mesmerized the residents. When some realized what the emptying of the bay meant, they ran for high ground, swiftly climbing the hill behind town. Whole families raced against death.

When the water poured back into the bay, it was rushing at fantastic speed, a wave about thirty-five feet high. It smashed into the village, crushing homes, drowning people, wiping out a way of life, and according to some estimates, continuing seventy to one hundred feet up the hillside. Everyone who had taken refuge in the church perished. Everyone who didn't make it to higher ground died.

One newspaper account said twenty-three lives were lost, one-third of the village population. Every structure except the school was destroyed. When the water retreated, there was almost nothing left of Chenega Bay except horror.

"We heard a terrible crack," survivor Dorene Eleshansky told the *Anchorage Daily News* twenty-five years later. "It was the church, I think. Tide rolled in over everything. It all just washed away."

The hopes and dreams of many families were carried away with it.

In 2006, John E. Smelcer gathered stories of survivors into a slender book titled *The Day That Cries Forever.* Those scattered Chenega citizens talked of their nightmares and their experiences. The book states that 120 people lived in Chenega at the time of the quake and that twenty-six died.

Although private homes did not have television, movies were screened at the school. A film showing was planned for later that evening after dinner. The school, the only building that survived intact, would have held a larger proportion of the population had the tidal wave struck later while villagers were watching *The House on Haunted Hill,* starring Vincent Price.

"Maybe a lot fewer people would have died then," Margaret Borodkin said.

Borodkin was lucky to have survived. When the earthquake struck, she was at home. While trying to protect herself by standing in a doorway, she was hit by a large object, possibly a piece of wood, which knocked her down. Unable to stand, she thought her leg was broken. Her mother was outside calling for help when the tsunami hit.

"After a few minutes I began to hear a loud sound," Borodkin told Smelcer. "It was really loud. I couldn't see because I was trapped on the floor, but it must have been the giant tidal wave. When it struck the house, it was as if a bomb blew up. Everything was crashing and rolling and smashing. I must have passed out during that because I don't remember anything else about that. I remember being scared. When I woke up I was out in the bay floating on a large piece of debris."

When Borodkin regained her senses, she realized the village had been leveled. She was wet, cold, and hurting. She started yelling for help. Someone on shore heard her cry and shouted her name, but no one found her until a fishing boat happened upon her amid floating debris. It negotiated a path to her, and strong hands belonging to Nick Elashansky and Marky Kompkoff lifted her out of the sea.

"They put three sleeping bags on me, but that didn't help," she said. "They later told me that I stopped breathing for five minutes, but that I eventually came to."

One of the fatalities was Borodkin's mother, Anna Vlasoff, who Borodkin last saw when she was out in front of their house, yelling for someone to help her injured daughter.

STEVE ELASHANSKY JR. was a five-year-old playing outside with his cousins. They were skipping rocks across the water but then switched to throwing stones at birds. Strangely, all four scored a hit on the same bird, and it fell to earth and died. To them, it was as if their misdeed had caused the earthquake that began rumbling minutes later.

"I thought God was mad at us for killing that bird," Elashansky said.

Caught unawares as the bay emptied and the tidal wave began, Elashansky was separated from his cousins and left alone until another resident called to him and got the boy to climb and crawl uphill to the man's house. From that perch, he watched in amazement as his community got swept away by the turbulent water.

"Below me I saw entire houses swallowed as the earth opened up to eat them," he said. "I could hear the sound of wood snapping and glass breaking as houses slid into the cracks, vanishing into the ground. I could see people running away from the waterfront. Behind them I saw a giant tidal wave, a wall of water almost a hundred feet tall. It was coming in fast. From where I was the people looked like little ants compared to the wave. I saw our house vanish under the first wave."

Elashansky did not realize at the time that his father and baby sister, inside the house, were killed. The wall of water he described was not truly a hundred feet high, but to a little boy, and anyone caught in it, it must have seemed so. It was big enough to destroy a village.

LARRY EVANOFF was attending boarding school in Wrangell when back at Chenega Bay his home was destroyed and his parents, who had sought shelter in the church, were killed. Larry didn't go to school for days, crying all of the time. Even nearly half a century later, he still sheds tears when he talks about the day when his life changed so dramatically.

Avis Kompkoff, who lost several relatives, had run up the hill, losing his slippers but continuing barefoot. The night of the quake he remembers staying awake, gathered around a fire. He was already mourning.

Part of that barefoot journey was in thick snow. It was certainly still winter at higher elevations. Mary Ann Kompkoff, one of those running for high ground, remembered the snow as an obstacle.

"We were running through deep snow," she said. "We kept falling in the snow, which slowed us down.

Survivors witnessed their community being wrecked and their relatives being killed by a powerful tsunami that picked the village clean—so neatly that when a Coast Guard plane flew overhead to survey Chenega Bay, it reported at first that things were okay. That false information was not corrected until Jim Osborne, the mail pilot, alerted authorities. Osborne began ferrying survivors to Cordova.

Unlike larger cities, Chenega was not immediately rebuilt. During the months following the quake, the surviving residents were relocated to such places as Tatitlek, where they lived in Army tents.

Residents soon dispersed to other communities, no longer neighbors, but acquaintances who shared a terrible experience and stayed in contact only intermittently. Each year on March 27, survivors returned to hold a ceremony and remember the dead, holding their own private Memorial Day service. For twenty years some of those people talked of rebuilding Chenega at the old site, but the idea was eventually abandoned because memories were too painful. A new Chenega was ultimately established on Evans Island, twenty miles from the original village.

Many of the initial survivors have since passed away, but children of Chenega who grew up minus parents and brothers and sisters still hold a place in their hearts for those who perished on Good Friday, 1964. They miss the way things used to be.

Donia Abbot told Smelcer that not only did many humans die that day, but that considerable spirit did as well for people who survived and were forever wounded by the end of a way of life.

"My people were forced to move away from the only home they had ever known," Abbot said.

IN FAIRBANKS, Rita Ramos took her two young daughters to church for Good Friday services late in the afternoon. There was standing room only and Rita stood, although she was seven months pregnant. When the church began shaking, the priest urged everyone to get outside.

"We watched the church steeple sway sideways," Ramos said.

Later when Ramos went shopping, store clerks and customers were listening to reports about the earthquake on radios.

THE LARGEST AND HARDEST-hit cities in southcentral Alaska—Anchorage, Kodiak, Seward, and Valdez—got most of the attention after the 1964 earthquake, but in big and small ways many other places were affected, too.

Seldovia was a small fishing village on the Kenai Peninsula overlooking Kachemak Bay. Commercial fishing was the community's reason for existence, and that is what brought Dana Stabenow's mother, Joan Barnes, to Seldovia from Cordova in 1960. Joan worked on the fish tender *Celtic,* and she and Dana spent much of the year living on the boat. The rest of the time, mostly in winter, they lived in a home owned by Nils and Nellie Pilskog.

With fishing on hold, winter was quiet time in the community of five hundred. In summer, the place was booming, with five canneries humming as fishing boats off-loaded their catch. At that time of year, Stabenow, who became a popular mystery suspense novelist, said, "The low-tide smell was enough to gag a maggot."

Seldovia was carved into a rocky coast and built on a two-mile-long boardwalk. It has been referred to as a community on stilts. During its heyday, Seldovia partook in the boom of the king crab fishery, but like Kodiak suffered when the species dwindled in the North Pacific. It became the lesser-known Kachemak Bay neighbor of Homer, the city that bills itself as being "At the End of the Road." That road is the Sterling Highway connecting Homer to Kenai, Soldotna, Seward, and Anchorage. Seldovia is more remote.

March 27, 1964, was Dana's twelfth birthday. Normally, Dana and her friends roamed the town, fishing, swimming, riding on homemade boats, climbing trees, picking salmonberries, and even bear-watching.

However, this day was more formally planned, with a handful of friends from sixth grade invited for cake and ice cream. Dana began unwrapping her presents at the kitchen table, including a dress for her Barbie doll that she remembers all too vividly as a peculiar outfit for an Alaskan Barbie. The gold sheath was not like anything she ever saw anyone wearing in Seldovia and, in fact, was not like the kind of dress she ever saw anyone wear in Alaska.

Suddenly one of Dana's friends, Katy Quijance, gave her a hard shot to the side. Although Dana tried to ignore her to concentrate on opening presents, she soon realized her friend was trying to tell her that a picture hanging on the wall across from them was swinging back and forth.

"I realized the whole house was moving back and forth, back and forth, just like the picture was," Dana said. "A second later the rest of the girls realized what was happening."

"Everyone stay right where you are!" Dana's mother told the children.

"We froze," Dana said. "The house kept moving, back and forth, back and forth."

After a couple of minutes, everyone thought the quake was over. One of the girls said, "Wow! That was a big one." It was, but they were mistaken. It wasn't over. The shaking started up again.

"This time it was a lot faster and up and down like a jackhammer," Dana said. "A curious thing about earthquakes is how loud they are, and this was deafening (with the) cupboard and the refrigerator opening, dishes and cans crashing to the floor, the house creaking and groaning, and the earth rumbling and tearing."

Everyone began shrieking and Dana's mother ordered everyone out of the house. Standing up was a challenge. They had to hold onto the chairs for balance, lean against the wall to make progress, and hold onto the porch outside. The spruce trees did that whipping back and forth dance, and the ground yawned open and slammed shut. Two latecomers to the party, who had been delayed by Easter choir practice, screamed as they were nearly swallowed by the ground.

Dana was less alarmed than thrilled by this fantastic event coming in the middle of her birthday party. It made the party seem so cool.

"I thought it was great," she said.

But the party was over. The girls went home. Dana wandered downtown to see what was going on. Goods had been thrown off the shelves at Bayview Mercantile. A radio news report indicated the airport control tower in Anchorage had collapsed.

"I began to realize that the earthquake hadn't been a fun thing put on for my personal enjoyment," Dana said.

An hour after she returned home, Seldovia's mayor knocked on the door to inform Dana and her mother that there was a tsunami warning and people were evacuating to the school gym, the biggest building in town and located on the highest ground. Everyone headed for the school except the fishermen, who took their boats out to try to ride things out on the water.

"It was dark when the tsunami hit," Dana said. "I was standing outside the gym, holding the hand of my mother's friend, Maka, listening to it—this continuous, menacing growl of moving water. It was very distinctive, unlike anything else, and I can still hear it."

The waves did not hit Seldovia as hard as they struck other communities, but the water came in high enough to soak the first floor of buildings and powerfully enough to cause serious damage in the new small-boat harbor. Seldovia residents felt lucky. However, when the next high tide came in April, it brought an unwelcome surprise. The earthquake had caused the land to drop about five feet, leaving Seldovia's boardwalk under water. In order to walk around town, residents had to wear rubber boots. It was impractical to cope with this inconvenience and an invitation to disaster if a rogue tide occurred or another tsunami.

Foot-high water seeped into homes and businesses, in some cases rendering first floors unusable. Eventually the boardwalk was relocated and rebuilt, and the old Russian Orthodox cemetery was moved to higher ground.

Over time Seldovia lost permanent population and became more of a tourist-oriented community than a commercial fishing hub.

WHITTIER, THE SMALL community that until a few years ago was reachable only from Anchorage by train or road running through a tunnel carved out of a mountain, was home to about seventy people in 1964. The town was hit hard by the quake, but there was not much damage. In *The Strangest Town in Alaska,* author Alan Taylor described how the earthquake felt, depending on where one was: "The ground shook jarringly for those who were on bedrock. Those who were on soil felt a round-and-round motion."

The epicenter was only sixty miles from Whittier in Prince William Sound. Buildings were severely damaged, and an old oil tank farm caught fire. Union Pacific Oil Company's tanks spilled thousands of gallons of oil. Thick, black smoke billowed skyward as the oil burned, some of it on the surface of the water.

"The sea halted its assault after three waves, leaving the town site's shore in a shambles," the book reported.

The tidal waves, created by landslides, killed thirteen people in Whittier.

ABOUT THIRTY-FIVE miles from downtown Anchorage is Girdwood, technically part of the municipality, although it has its own identity. Ten miles farther south on the Seward Highway is the community of Portage.

Approaching Girdwood today, southbound traffic passes stunted, gray, naked trees on the right, seemingly relics from a long-ago forest fire. Actually they are dead remnants of the earthquake, when the land shifted and salt water from Turnagain Arm rushed in and killed their roots.

Portage was home to about seventy-five people in 1964, but today it is known almost solely as the site of Portage Glacier. The buildings of a half-century ago were wrecked. Years later, one former resident, Bill Glasshof, told the

Anchorage Times that he was working in his shop when the Big One struck, overturning Glasshof's work bench and tossing around his tools.

The sharp sounds of trees snapping unnerved a co-worker, who Glasshof said was spooked by the likelihood of a Russian invasion.

"The Russians are coming. The Russians are coming", the man warned, as if he were advertising the Alan Arkin movie of the same name.

THE GOOD FRIDAY Earthquake, rated one hundred times more powerful than the atomic bomb dropped on Hiroshima, was felt as far away as Louisiana. "Noticed in Louisiana" would be a more accurate description. On the West Coast, however, the impact was destructive.

The tsunami feared in Anchorage struck farther south. In Depoe Bay, Oregon, the self-described "smallest harbor in the world," a family of four camped at Beverly Beach State Park was killed, crushed by logs or drowned. Although other cities on the Oregon coast experienced some high water, damage was limited and there were no other fatalities in the state.

The impact was worse in Crescent City, California, at the end of a half-moon-shaped bay, where a series of four large waves early the morning of March 28 destroyed the boat harbor, wrecked a sawmill, tossed giant redwood trees around like toothpicks, and inundated twenty-nine city blocks, killing eleven people.

In a headline, the *Humboldt Times* proclaimed:

TITLE WAVE LEAVES WAKE
OF DEATH AND DESTRUCTION

Clarence and Peggy Coons, resident lighthouse keepers on Battery Island, about three hundred yards offshore, watched in awe as five gigantic waves clobbered the community. Peggy Coons wrote a compelling, first-person account of what they saw on that grim day.

Peggy had gotten up in the middle of the full-moon night to go to the bathroom but sensed something was amiss with the water. She was used to seeing hundreds of rocks above the water line, but none were visible.

"I was becoming alarmed," she wrote. "The ocean seemed to be rising around us and here we were alone on this tiny island." Coons woke up her husband. They dressed, put on jackets, and went outside. "The air had an odd stillness as we went about looking in every direction. The night was unusually bright and the air was very still. There were no sounds, not even water lapping at our rock."

When the Coons saw a large wave approaching, they moved to higher ground on the island facing Crescent City. When Peggy saw how big it was, she yelled, "Oh, my God, no, it will flood the town!"

And it did.

They could see buildings falling and hear the sound of wood cracking and glass breaking, and then they watched the tide recede and carry pieces of the community into the harbor. When the second wave hit, the Coons wondered if the one-hundred-year-old lighthouse, their home for two years, would hold up.

"We stood glued to the spots where we were," Peggy wrote. "We were too frightened to move so I don't know how long we stood there."

On the third wave, even bigger than the others, the lights on Highway 101 went out, and fires broken, out on the south end of town. The Coons checked the lighthouse. Their island was still safe. Not so the town as a fourth wave massed, at first emptying the bay into "a black canyon," and growing in force ready to strike again. Nothing seemed as ominous as this bay suddenly empty of water under a full moon. The couple embraced, fearing for their lives.

"Suddenly, there it was," Coons said. "A gargantuan wall of water was barreling right toward us. It was terrifying. This wet, mountainous mass of destruction was looking for a place to happen. It stretched from the floor of the ocean to high in the sky."

The Coons rushed to climb the lighthouse tower, but they could not move fast enough. The wave hit the island, but miraculously it was diverted around the protuberance of land and didn't even wet them. The fifth wave was smaller. Then the tidal waves ceased.

Crescent City was burning on one end, under water in many areas, and disintegrating at its core. Hunks of buildings, cars, furniture, clothing, everything and anything, were sucked out to sea. Surrounding the Coons was "damage, damage, damage." Throughout it all, however, the lighthouse beacon kept shining.

The series of tsunamis that found land in Crescent City, California, had traveled about eighteen hundred miles from Prince William Sound to do its nasty work. The Good Friday Earthquake was spreading heartbreak far beyond Alaska.

It was soon discovered that the 8,700-mile-long Commonwealth telecommunications cable, located seven and a half miles off the shore of Canada that linked Great Britain with Australia and New Zealand, was broken by the earthquake.

By coincidence, the Seismological Society of America was conducting its annual meeting at the University of Washington. One scientist was dining at the top of the Space Needle, 590 feet above the ground, when he and his party felt the rotating restaurant take a hit. The restaurant stopped turning.

One of the questions discussed by the seismologists in Seattle that Good Friday, but to which no definitive conclusion was reached, was whether scientists could predict earthquakes with any certainty. The answer on that fateful day was still no.

An iconic image from the 1964 earthquake was this scene, top, from 4th Avenue in downtown Anchorage, where stores settled eleven feet below street level partly in response to violent horizontal movement. The marquee of the Denali Theater, bottom, came to rest on the sidewalk on 4th Avenue.

The damaged J.C. Penny store in downtown Anchorage became an iconic image of death and destruction.

A "graben," a depressed block of the earth's crust bordered by parallel faults, was formed along the head of L Street in Anchorage. The trough sank seven to ten feet, undercutting and tilting homes.

These ruined homes in Anchorage's Turnagain by the Sea neighborhood testify to the destructive power of the earthquake and resulting landslide.

This house destroyed in the hard-hit Turnagain area of Anchorage was typical of many that fell into fissures caused by the quake and by a massive landslide that sent homes sliding toward Cook Inlet.

Seen here climbing through earthquake rubble, rescue teams comprised of civilians and military personnel conducted a house-to-house search for survivors in the Turnagain neighborhood.

Governor Bill Egan inspects damage in downtown Anchorage in an Alaska State Police cruiser, top. Within 90 days, Alaska had received $17 million from President Lyndon Johnson's Disaster Fund. Bottom: cleanup got underway quick on 4th Avenue.

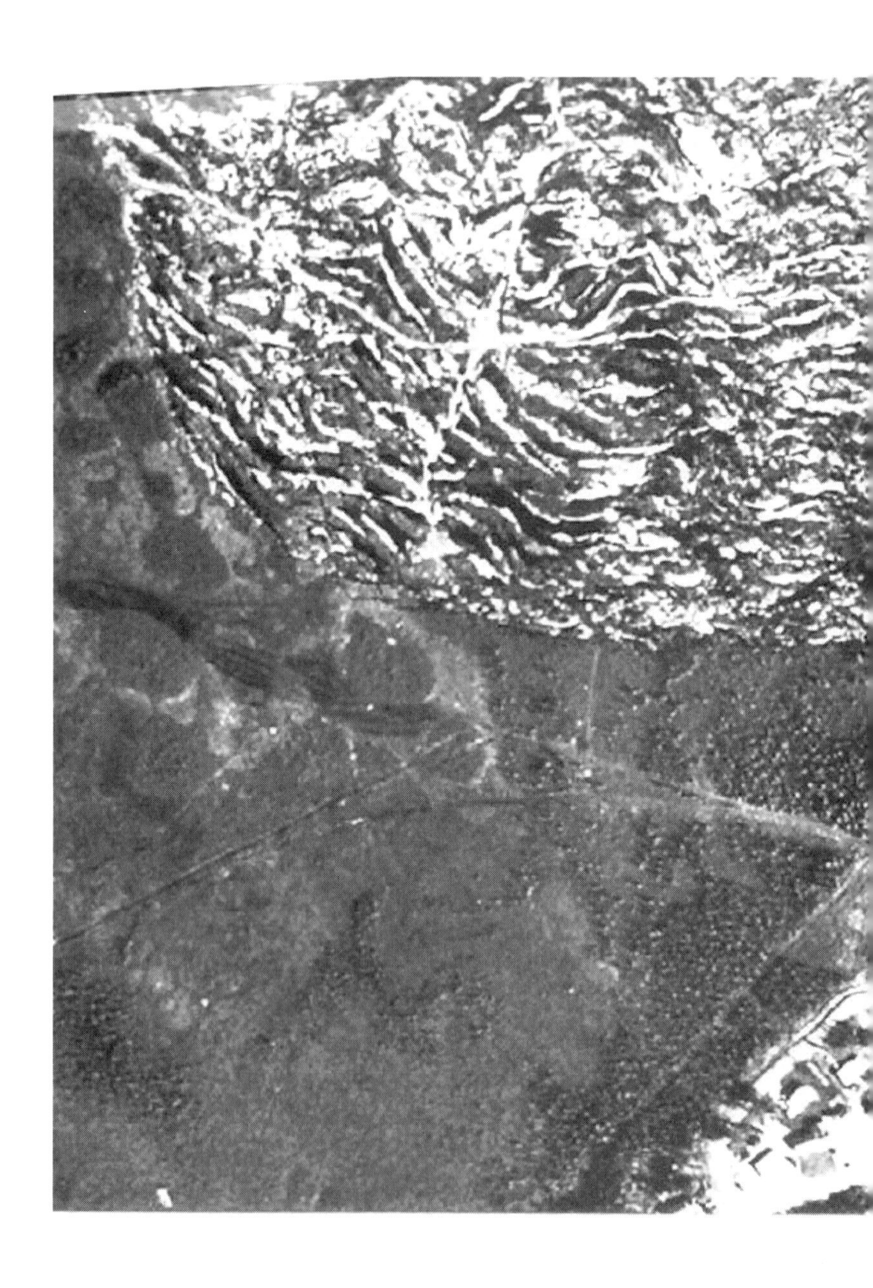

This aerial view is a dramatic illustration of where a section of the Turnagain Bluff slid into Cook Inlet, taking homes with it. The soft clay bluffs dropped due to soil liquefaction, caused when loosely packed, water-logged sediments shake and lose their cohesive strength.

The Alaska Native Medical Center in Anchorage narrowly escaped this landslide. Built in 1953 as a tuberculosis sanitarium, the heavily damaged structure was replaced by a new facility in May 1997 six miles away.

In Valdez, an abandoned car is wedged in a large crack as oil tanks burn in the background. The quake caused many roads in town to split apart as millions of cubic yards of earth-to-slide into Valdez Bay.

A failure of oil-storage tanks at the Union Oil tank farm in Valdez triggered a massive fire that destroyed what was left of the waterfront and burned for two weeks.

Landslides of glacial deposits, top, destroyed almost a mile of Alaska Railroad track on Potter Hill near Anchorage. Bottom: on Turnagain Arm, rails were torn from their ties and a railroad bridge was buckled.

The Twentymile River Bridge on the Seward Highway, top, broke into pieces and fell into the river. About twenty miles of the highway were destroyed, too, including a section split by fissures, bottom.

Soldiers from Fort Wainwright near Fairbanks cleared debris from the dock area in Valdez two days after the earthquake. The military was praised for its quick response to requests for disaster assistance.

This view of downtown Kodiak, a thriving commercial fishing port on Kodiak Island, shows debris and damage from a combination of the powerful earthquake and tsunamis that followed.

Misplaced fishing boats testify to the enormous power of the 1964 earthquake and tsunamis that dealt Kodiak a one-two blow.

A tsunami devastated Seward, top, where twelve people died. The giant wave roared in at an estimated four hundred miles per hour, creating a tide one-hundred feet above normal. A boat was left high and dry, bottom.

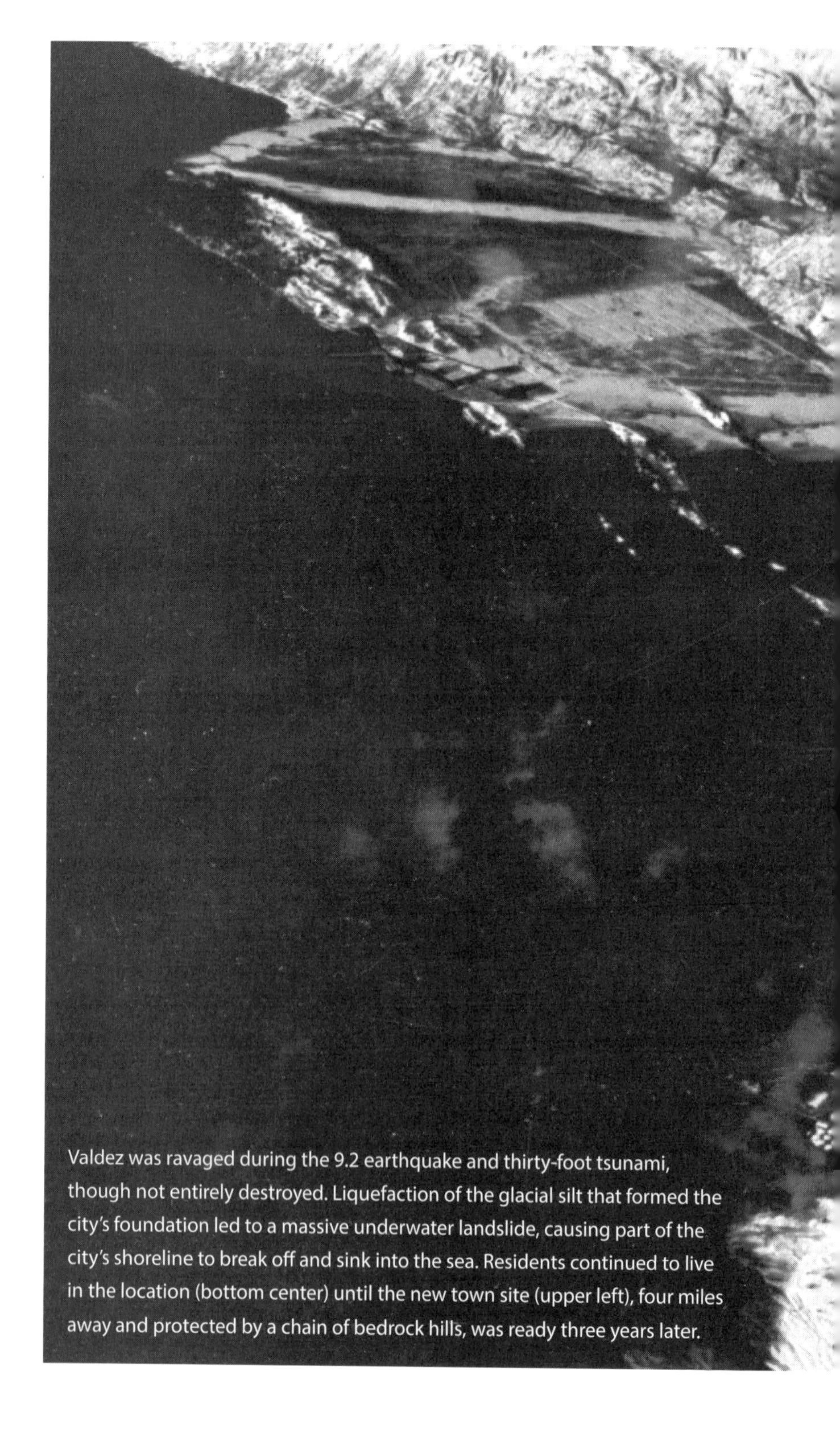

Valdez was ravaged during the 9.2 earthquake and thirty-foot tsunami, though not entirely destroyed. Liquefaction of the glacial silt that formed the city's foundation led to a massive underwater landslide, causing part of the city's shoreline to break off and sink into the sea. Residents continued to live in the location (bottom center) until the new town site (upper left), four miles away and protected by a chain of bedrock hills, was ready three years later.

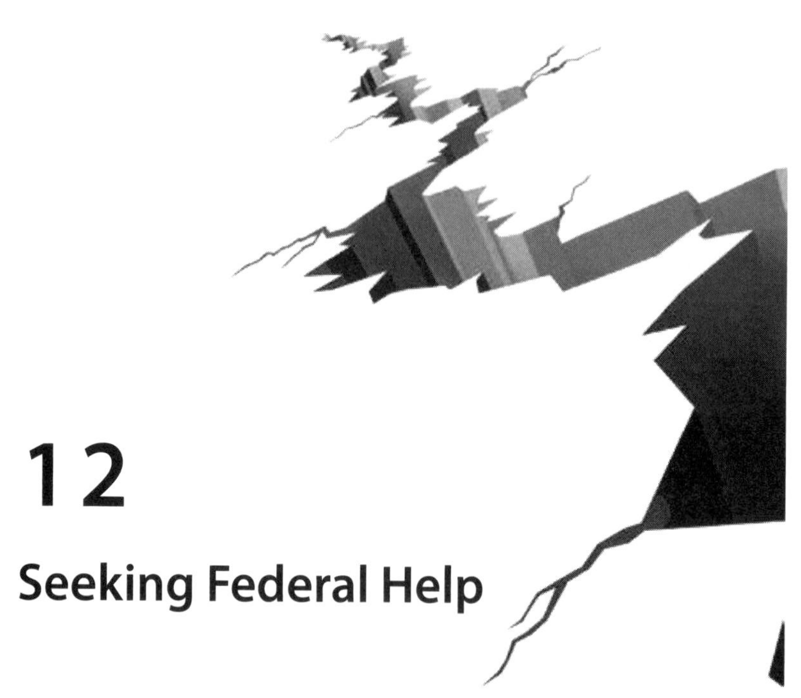

12

Seeking Federal Help

The first estimate of damage to Anchorage was $54 million, and $350 million statewide, yet within days that was increased to an estimated $500 million worth of damage statewide. In Anchorage a quick analysis by authorities also indicated that $100 million was lost in the local tax base due to residential and commercial property losses.

Besides the loss of private homes, apartment buildings, and hotels, all public utilities delivering electricity, telephone service and water, were damaged and temporarily put out of commission. The Alaska Railroad was shut down. The tracks had been rearranged so violently no train could ride on them.

Before any rebuilding took place, Governor Egan and Anchorage Mayor Sharrock wanted the U.S. Army Corps of Engineers to analyze the ground to make sure stable locations were chosen. No one wanted contractors to throw up new buildings in the most earthquake-prone zones. Although aftershocks still rippled through the region, no one thought that there ever would be a repeat. Scientists considered the Good Friday Earthquake to be an occurrence that would take place once every three hundred years.

Sharrock announced quickly that Anchorage was determined to rebuild as good as new. However, he did discuss the possibility that the downtown business area, Fourth Avenue in particular, which had been torn to bits, could be relocated.

Sharrock "would not discount" the idea of moving the core area of downtown. Turnagain, of course, had been hit even harder, with houses knocked off their foundations and then crumbling as they tipped downhill towards Cook Inlet. Perhaps aware that many of the richest and most powerful people in Anchorage maintained homes in that area, Sharrock issued no quick declarations that they would not be allowed to rebuild there.

"When we know where the safe areas are we can plan the city," Sharrock. said "It has set us back a little—and I don't think we will be affected long. There is no need to feel that this place is any more dangerous …"

Meanwhile, Alaska politicians were pressuring the federal government to act quickly. Senator Ernest Gruening labeled the event as "the largest disaster ever to strike a state in the history of the United States. Considering the total population and resources of the state, the earthquake struck at the heart of Alaska." Gruening said that after all of the foreign aid spent to assist other countries in need over the years, he felt this was "Congress' hour of truth on whether it will treat our own people in their need as they have given indiscriminately to all peoples around the world."

Gruening, Senator Bob Bartlett, and Edward McDermott, the president's envoy, were dispatched by Lyndon B. Johnson in a presidential jet to Alaska to gather information and make a report. In his autobiography *Many Battles*, Gruening, a Democrat who had been Alaska's territorial governor from 1939 to 1953 and served in the U.S. Senate from 1959 to 1969, wrote that he had inspected five cities, including Anchorage, and he was awed by the destruction.

"It was hard to believe the evidence of our own eyes," said Gruening, who passed away in 1974. "Great buildings had collapsed. Private houses had been demolished or were badly askew. In Anchorage's fashionable residential section Turnagain by the Sea, not only had a score of houses on the bluff been swept out to sea in fragments, but the ground on which they had stood had been washed out with them."

As Gruening flew overhead inspecting burning oil tankers in Seward, the price of reconstruction mounted in his mind until he settled on a figure of $750 million. Gruening said $500 million was the minimum needed. By April 5, Egan was agreeing with Gruening on the larger figure, and when he landed at Andrews Air Force Base before meeting with President Johnson and his experts on disaster relief, that's what he told reporters. Sometime after that declaration, the figure of need dropped back to $500 million by other sources who had more time to make calculations than Gruening did.

By the time Egan flew to Washington to plead for help in person, he was told Congress had voted unanimously to create an earthquake fund to assist Alaska and that $50 million had been approved. He welcomed that good news, but said it was not enough to get the job done. The initial dollar figure was "a drop in the bucket compared to the overall losses," he said.

Egan said flatly that federal grants were needed.

"It is absolutely necessary that this money come from Congress," Egan said.

There was no chance that all of the cash was going to come from the federal government with no strings, or loans attached, and on April 1 Egan also asked the state legislature to approve a $50 million bond issue for the state to borrow money to pay for the cost of replacing homes, businesses and major industries wrecked or damaged by the earthquake.

One reason federal aid was so important is that a very small percentage of Alaskans owned earthquake insurance at the time, perhaps amounting to coverage of 1 percent of $500 million worth of losses.

Egan jumped into an intense round of meetings in Washington with Alaska leaders, congressional leaders, and administration leaders, and he returned to Alaska buoyed by the positive reception he received and convinced the federal government understood the magnitude of the reconstruction effort.

"I believe it is the attitude of Congress and the American people that there will be a program to insure that all those who wish to rebuild can rebuild without outlandish assessments," Egan said.

Senate Bill 2881, a group of amendments, was actually introduced on May 28 by Bartlett, another Democrat who represented Alaska in the Senate between 1959 and 1968, with Gruening and Washington State's senators as co-sponsors, and it was called the "Alaska Omnibus Act." In part it read, "The Congress recognizes that the State of Alaska has experienced extensive property loss and damage as a result of the earthquake of March 27, 1964, and subsequent seismic waves, and declares the need for special measures to aid and accelerate the State's efforts in providing for the reconstruction of the areas in the State devastated by this natural disaster."

Although the sentiment for helping Alaska was there, Gruening quickly ran into a political roadblock in Washington. At the time, foreign borrowers had only to repay loans at a rate of three-fourths of 1 percent. The Small Business Administration wanted Alaskans to pay its customary 3 percent rate interest on disaster loans.

In the *Congressional Record*, Gruening declared that the rate differential "may well be the difference between recovery and failure" for Alaskan companies. Gruening did not prevail with his entire package of requests, although Congress approved funds for urban renewal requiring only 10 percent matching funds from state and local governments.

In reality, while Gruening had hoped for more, he warned Egan early on that most of the assistance for private citizens might not be direct payouts.

While officials traveled from Washington, D.C., to Alaska to inspect damage, Egan sought to keep calm and provide information and reassurance. His basic form of contact was the radio. In some ways, during the crisis, Egan's reports served the same type of role as President Franklin Delano Roosevelt's fireside chats to worried Americans between 1933 and 1944 during the Great Depression and World War II.

As part of his statewide radio address, Egan mentioned, "money for private rebuilding will not—in most cases—be outright grants." He was reading the tea leaves properly on how Washington was leaning. Egan's talk led the Anchorage City Council to adopt a resolution asking for federal aid for public utilities, roads, and other public buildings.

Americans were inspired to help the forty-ninth state when they read their home newspapers. The headlines were huge.

The *San Francisco Examiner* trumpeted:

QUAKE DISASTER

The *San Francisco Chronicle* reported:

TERROR IN ALASKA

The *Los Angeles Times* published a rare "extra" edition with a simple but huge headline:

ALASKA QUAKE

The news stories reported that the quake was so powerful the needle indicators measuring earthquake magnitude had swung so forcefully that they bounced at the end of the scale.

Donations poured into Alaska from a variety of sources.

A man from St. Louis named Gordon Coates and his wife Thelma had come to Alaska as tourists in 1963, and the place rubbed off on him. He contributed $2,000 to emergency relief, $1,000 to the Salvation Army and $1,000 to the American Red Cross. He mailed the checks to the *Ketchikan Daily News*. In a note he said in part, "I certainly was shocked when I heard of the earthquake in Alaska. I believe a person feels this more when you have seen the area and visited some of the places involved."

A representative of the Japanese government presented Egan with a $10,000 check for Alaska's reconstruction fund. Nationwide five thousand Jaycees chapters were asked to raise money for Alaska by their Anchorage brethren. By early April, Egan also had a check in hand for $10,000 from Pope Paul VI. Right about that time, the executive board of the International Union of Operating Engineers pledged a $15,000 gift to Alaska for rebuilding to be administered through the Salvation Army.

Although it was obvious that there would be a tremendous amount of construction work and building going on in Alaska, the state put out a "Not Welcome" mat, saying that people should not come north just to make money and that Alaska had all of the workers it needed. Alaska was committed to rebuilding itself.

13

What Happened

A half-century after the Good Friday Earthquake, it remained the second-largest earthquake in history, with a magnitude of 9.2. The world's largest quake, with a magnitude of 9.5, occurred in Valdivia, Chile, on May 22, 1960.

Forty years after the Alaska earthquake, the Dec. 26, 2004 earthquake in the Indian Ocean near Indonesia measured 9.1. Two others, a Nov. 4, 1952, earthquake in the Kamchatka region of Russia and the March 11, 2011, earthquake in Japan, both measured 9.0.

Alaska is located in what is called the Pacific Rim Ring of Fire, a region known for its volatile nature under the earth's surface. Earthquakes are much more common in this area of the Pacific Ocean than in other parts of the world.

The Good Friday Earthquake was caused by two large plates of the earth's surface shifting. The plate tectonics, as it is called, consisted of the Pacific Plate being pushed under the North American Plate. This emission of immensely powerful force was felt by people at first through small tremors that grew increasingly intense and lengthy. People in Anchorage felt the earthquake for five and a half minutes, while those located elsewhere reported feeling the quake for between four and five minutes.

Beneath the ocean, layers of soil became liquefied and shifted abruptly. This provoked landslides that slid into waters near Valdez and Seward, causing tidal waves. The epicenter of the quake was in Prince William Sound. Fifty miles distant, Valdez was the closest of the significantly damaged Alaska communities.

Although earthquakes are recorded somewhere in Alaska's 586,000 square miles every day and those strong enough to be felt are commonplace, the magnitude of the Good Friday Earthquake was such an aberration, it is regarded as an event of nature that will occur only once every three hundred years.

Scientists believe the Pacific Plate rotates between five and seven centimeters a year, compressing and warping the earth's crust in the southern regions of Alaska. When the compression is relieved by land moving back and forth over the plate, an earthquake occurs. A study undertaken by the National Science Foundation and Office of Naval Research written by George Pararas-Carayannis indicated that "the net horizontal movement of the Pacific Plate under the North American Plate was about nine meters on the average in a southeast direction." That land did not move gently. The Latouche Island area of Alaska moved eighteen meters.

According to the Seismology Department at the University of Alaska Fairbanks, the state experiences 50 to 100 quakes a day, 400 to 700 a week, or about 24,000 earthquakes a year. According to a 2006 study, three of the ten largest earthquakes in the world and ten of the fifteen most powerful quakes ever recorded in the United States took place in Alaska. The research suggested Alaska could expect one earthquake of at least 8.0 magnitude every thirteen years and one of at least 7.8 every year. Any mention of 9.0 was ignored because it was off the charts.

When it comes to the land rocking and rolling, it is difficult to beat Alaska. Yet most of the quakes occur in uninhabited areas and cause little if any property damage.

The earth contains twelve plates, and they, part of the planet's "upper mantle." The earth's crust is the top layer of ground. When an earthquake occurs, depending on how powerful it is, its location, and depth, the force can raise the land and create new mountain ranges. The earth's crust extends less than twenty miles deep, and the upper mantle is an additional five hundred feet deep. Plate tectonic action takes place in the upper mantle.

Another by-product of the land being rearranged is where it may drop. The earthquake and tidal waves that battered Valdez lowered the bottom of the harbor between 50 and 150 feet, studies showed, and in Seward the harbor dropped to nearly 450 feet from a low point of 120 feet before the quake. There were fifty-two aftershocks that were notable in size, with eleven of those more than 6.0 in the first three days. The largest aftershock was 6.7.

Initially, the Alaska earthquake was estimated at 8.4 on the Richter scale but was upgraded later, first to 8.6 and then to 9.2.

As late as the early 1900s, scientists had little understanding of what caused earthquakes. Some believed they stemmed from underground explosions or something of that nature. The first measurement of the power of earthquakes that could be accurately applied was developed in 1935 by Charles Richter, a professor at the California Institute of Technology.

Richter called his invention "a magnitude scale." In an era before computers, Richter's handwritten graphs and charts were a painstakingly slow trial-and-error way to make sense of seismic events. Richter was able to apply a logarithm to narrow the range of numbers he worked with. A key element of factoring was distance from the apparent center of the shock.

The way the scale was developed, with tenths separating whole numbers—5.2, 6.3—many people were fooled into thinking there was little difference in measurements so tightly packed together. However, magnitudes are tremendously more powerful within the small range, so a 4.0 earthquake is "30 times more energetic than a magnitude 3," according to Susan Elizabeth Haugh, Richter's biographer. "A magnitude 5.0 is one thousand times more energetic than a magnitude 3.0. Energy increases by a factor of thirty rather than ten for arcane reasons related to the nature of the earthquake process."

A *U.S. News and World Report* story of April 13, 1964, referring to scientific expertise, called the energy released by the Good Friday Earthquake "about a million times as great as that released by the Hiroshima and Nagaski atomic bombs."

For decades, all references to the size of earthquakes were calculated by the Richter scale. In recent years, scientists have depended more on the "Moment

Magnitude scale," developed in the 1970s. However, in public vernacular, the Richter scale is still better known and more often used.

Although some people believe that the Richter scale is a one-to-ten scale, it is not. Charles Richter did not have a cap on the scale; it was open-ended. The Moment Magnitude scale is regarded as providing a more accurate reading on the largest earthquakes in particular. The general description now applied to the Good Friday Earthquake is that it registered a 9.2 magnitude.

While the magnitude of the earthquake was revised upward by scientists, the cost of the damage from the initial estimates of $500 million to $750 million was revised downward to $300 million.

Besides the shaking apart of Anchorage's streets, the earthquake caused physical changes in certain parts of the landscape. Parts of the ocean floor between Kodiak and Montague Island were lifted fifteen feet. Latouche Island moved twenty feet closer to Montague Island, and mountains on the Kenai Peninsula moved five feet.

The 1960 Chilean earthquake, the only one greater than the Alaska earthquake, was far more devastating because it touched more populated areas. The death toll from the Alaska quake was 131, with 115 of those people perishing in Alaska and the other 16 from tidal waves on the Pacific Coast. In Chile, more than 2,000 people died and more than 5,000 were injured and 50,000 homes were wrecked. Chile also experienced tidal waves as high as 24 feet.

Impact from the Chilean tsunami was ruinous elsewhere, too. The water rushed to Hawaii and across the Pacific to Japan, where 150 people died from the waves.

Eyewitness accounts of the size of tsunamis hitting Alaska vary widely after the Alaska earthquake. Scientific analysis indicated that the tidal wave that wiped out Chenega Bay was 89 feet high; the largest of the three that smashed into Whittier was 42 feet high; waves hitting Seward and Kodiak were 30 feet high; and in remote uninhabited Shoup Bay, about eight miles north of Valdez, the wave was 200 feet high.

Tsunami effects were registered in twenty-five Alaska communities, from the worst-hit places of Chenega Bay, Valdez, Whittier, Seward, Kodiak, and Kalsin Bay to Sitka, hundreds of miles away in southeast Alaska, where a dock collapsed but no one was hurt.

Tsunamis of various sizes were recorded in more than 240 places in Alaska, Canada, Washington, Oregon, California Hawaii, Mexico, El Salvador, Costa Rica, Colombia, Peru, Chile, Pago Pago, Mariana Islands, Wake Island, Australia, Japan, the Marshall Islands, New Zealand, Kamchatka, and Antarctica.

Not mentioned was Texas, where in the immediate aftermath of the earthquake, the Associated Press in Houston reported unusual tide variations and mysteriously bobbing boats in the state's harbors. In Port Arthur, the tide dropped between six and seven feet, the AP said, with "a loaded grain ship (that) bobbed up and down like a cork six or seven times." Abnormal waves were reported elsewhere in the state, and workers on oil rigs in the Gulf of Mexico said they saw a five-foot drop in the tide. In New Orleans, some boats tied up were heaved over in six-foot waves rushing through rivers and bayous.

"The water rose about six feet above normal all at once," a watchman on the New Orleans Industrial Canal told the AP. "It was one of the wildest scenes I've seen in a long time. The water was rolling, barges began to move in and out, and the lines began to turn and break."

New Orleans is more than 3,400 miles from Anchorage.

ROGER HANSEN, the Alaska state seismologist stationed at the Alaska Earthquake Information Center in Fairbanks, once said, "We are the exporters of tsunamis. Alaska is more likely to be damaged from a local tsunami, but any time we have one of those we're sending a huge wave to Hawaii or California."

As a result of the 1964 earthquake, the West Coast and Alaska Tsunami Warning Center was built in Palmer, Alaska, forty or so miles north of Anchorage, in 1967.

In a paper written for the center, Thomas J. Sokolowski, who called the Shoup Bay wave the highest recorded during the 1964 earthquake, said that "more than 90 percent of the deaths in Alaska during the 1964 earthquake and subsequent tsunamis were due to the tsunamis. The potential death and destruction from tsunamis make the coastal area of Alaska extremely dangerous and necessitate continuous, 24-hour earthquake monitoring for each day of the year by a full-time staff at the West Coast and Alaska Tsunami Warning Center."

In the years following its inception, the center did not always have full-time staffing, although a rule required that employees had to live nearby so they could get to the center quickly in case of emergency. In 2005, staffing boosted to a level sufficient to monitor activity seven days a week, twenty-four hours a day. This followed the massive tsunami that struck Christmas week of 2004 in Asia.

That year, the intense earthquake in the Indian Ocean brought worldwide tsunami awareness to a peak as waves crashed into Sumatra and Thailand and sent smaller ripples across the sea. In Alaska, Dutch Harbor experienced a five-inch wave, but it could have been much worse.

Almost immediately after the earth stopped shaking, scientists and government officials began to analyze what happened, assess the impact, imagine the future if another earthquake of similar magnitude hit Alaska or elsewhere, and sought to develop methods that would pinpoint more accurately when a quake would hit. To date, the latter remains a very iffy science with little way to warn people early enough for careful, organized evacuations.

One study of what occurred on March 27, 1964, in Alaska was titled, "The Great Alaska Earthquake of 1964." It was reported by "The Committee on the Alaska Earthquake of the Division of Earth Sciences of the National Research Council at the National Academy of Sciences in Washington, D.C." The 1,200-page document was dated 1973 and contained a preface by Konrad B. Krauskopf of Stanford University.

Another report called "The Alaska Earthquake Effects on Communities," published by the U.S. Geological Survey, was penned by several authors, including Wallace R. Hansen, who studied Anchorage.

It has been stated repeatedly that the earthquake lasted five and a half minutes in Anchorage, but the USGS document left some doubt about the exact duration. "The duration of the earthquake at Anchorage can only be surmised owing to the lack of a strong-motion seismograph records," it said. "Although seismographs have since been installed, none was present in southern Alaska at the time of the quake. Intense seismic motions seem to have lasted three to four minutes, possibly longer." The report cited comments made by individuals who timed the earthquake on their wristwatches as having lasted between four and seven minutes.

The earthquake moved east to west into Anchorage at first, but the report indicates that as "the shaking continued, however, the motion is said to have shifted north-south and tall buildings that had first rocked to-and-fro east and west began to rock north and south, as well, in a complex combination of movements."

There were wild fluctuations in the estimates of property damage to Anchorage, with differences of hundreds of millions of dollars. "Total earthquake damage to property in the Anchorage area … perhaps will never be fully known," the report said.

Many more people were killed in outlying communities because of tidal waves, but Anchorage "because of its much greater size, bore the brunt of the property damage and property losses reportedly were greater there than in all the rest of Alaska combined," the report stated.

One observation was that the almost instant shutdown of power throughout Anchorage was probably a good thing for the community because "untold numbers of fires were probably avoided because of the lack of electric current in all the severed wires—and at a time, too, when water was unavailable for fighting fire."

Individuals in Anchorage and all over the region were startled when the earth opened up, mimicking the type of fissures normally seen on glaciers. The report relating to Anchorage said there were surely many more deep cracks than people noticed. They might have been hidden by forest and woodland, or located on the outskirts of town and hidden by snow cover.

"Many unobserved cracks probably formed in the less-accessible, snow-covered, undeveloped areas," the report said. "Frozen muskeg in and bordering swamps, for example, was very susceptible to cracking."

In the hardest-hit area of Turnagain Heights, 8,600 feet of bluff was reshaped by the quake with the ground at its forward-most point moving 1,200 feet forward toward the water, and the average advancement was 500 feet. The neighborhood was built on silty, clay-like ground and was not fortified by any natural barrier to prevent houses from falling down the hill after being rocked off their foundations.

Although there were no casualties tied to damages sustained by the Alaska Railroad, the railroad's damage was substantial, affecting bridges and buildings to the tune of more than $2.3 million.

Despite the loss of the Anchorage airport control tower, most of the rest of the airport's basic infrastructure was intact. No planes could fly into or out of

Anchorage International Airport immediately following the quake, but flights were resumed quickly. Initially air-traffic control was guided from a parked airplane, and then control was transferred to a tower at Lake Hood.

Anchorage was home to an unusually large number of geology specialists. "Within hours after the quake," the report stated, "many of these individuals had begun independent investigations of the location, severity, nature, and causes of earthquake damage."

A companion report about the effects of the quake on Valdez was compiled by Henry W. Coulter of the U.S. Geological Survey and Ralph R. Migliaccio of the Alaska Department of Highways. It referred to the tidal waves as stemming from "submarine slides." The report relied on an eyewitness account of Valdez geologist Charles H. Clark.

"The first tremors were hard enough to stop a moving person and shock waves were immediately noticeable on the surface of the ground," Clark said. "These shock waves continued with a rather long frequency which gave the observer an impression of a rolling feeling rather than abrupt, hard jolts." Clark noted that when cracks in the sidewalks and roads occurred, creating crevasse-like gaps, water shot into the air.

As everyone knew, the real damage was caused by the tidal wave caused by landslides from the 3,000- to 5,000-foot mountains surrounding Valdez. "The slide occurred so rapidly and with such violence," the report indicated, "that eyewitnesses on the shore were overwhelmed. Most of them have very little recollection of what actually occurred directly beneath their feet."

This report also relied on the views of *Chena* Captain M. D. Stewart, who said, "There were very heavy shocks about every half minute. Mounds of water were hitting at us from all directions. The Valdez piers started to collapse right away. There was a tremendous noise. I could see the land jumping and leaping in a terrible turmoil. We were inside of where the dock had been. There was no water under the *Chena* for a brief interval. The stern was sitting in broken piling, rocks and mud."

Stewart said it took about four minutes to get the boat going in the mud, and the ship couldn't turn. "We were moving along the shore, with the stern in the mud. Big mounds of water came up and flattened out. A big gush of water came off the beach, hit the bow, and swung her out about 10 degrees. If that hadn't happened, we would have stayed there with the bow jammed to a mud bank and provided a new dock for the town of Valdez."

The military evaluated its involvement in a study called "Operation Helping Hand: The United States Army and the Alaskan Earthquake, 27 March to 7 May, 1964." Credit for authorship was given to Major General Ned D. Moore, Commanding General United States Army Alaska, or USARAL.

At the Army Operations Center on Fort Richardson, Lieutenant Colonel Alfred H. Parthum Jr. and Master Sergeant Ralph C. Dahlgren were sitting down to dinner in "the war room" when the earthquake hit on March 27, 1964.

"My meal was still in the tray when I went off duty the next morning," Dahlgren said in the Army report.

Parthum, who had never experienced an earthquake, said he didn't know what was going on, and his first reaction was "surprise and bewilderment at the wave-like motion."

Parthum's first move was to try to reach the Yukon Command, but the commercial telephone lines were down. He also ordered an inspection of the post for damage. Without being summoned, officers reported to the Operations Center.

Although none of the quake impact was sufficient enough to sidetrack the Army from its mission and mobilizing troops to provide security in Anchorage and elsewhere, Fort Richardson did sustain $17 million in damage. The Alaska Command, which was split into USARAL, Alaskan Air Command, and Alaska Sea Frontier, was in steady communication with Governor Bill Egan and Anchorage Mayor George Sharrock.

The military response was swift in Anchorage. Troops helped protect the downtown area from looters and warned off people who might be in danger if a significant aftershock caused weakened buildings to collapse. Assistance was direly needed in outlying areas, too, but the Army had to obtain reports on the status of airports before it would airlift troops and emergency supplies into Seward, Valdez, and other communities.

Shipments began the next day with the arrival of daylight and continued for twenty-one days. Military Air Transport Service flew more than 2.7 million pounds of supplies into hard-hit areas of Alaska as part of Operation Helping Hand.

In addition, the U.S. military employed planes in the first two weeks following the disaster to survey the quake and tsunami damage from overhead in hard-to-reach areas, flying sixty-four hours on twenty-three flights to Seward, Whittier, Valdez, Kodiak, Seldovia, Cordova and north of Anchorage. Some 2,400 pictures were taken by April 10, showing damage along the Alaska Railroad and the Seward Highway, among other things.

There was plenty to see following the railroad line, which had considerable track bent and mutilated on the 112-mile route from Anchorage to Seward, on the 12.4-mile line to Whittier from the Seward Highway in Portage, and for about one hundred miles north of Anchorage.

Quick assessments of the track damage led to resumption of partial service between Palmer and Anchorage by April 7 and between Anchorage and Whittier by April 20. The latter seemed to be a particularly notable achievement, given that a wharf, transit shed, rail-barge slip, passenger depot, and marshaling yard, as well as track, all were destroyed.

One of the first services provided by the Army to the population was the contribution of fresh water. Six trucks with water trailers commuted between Fort Richardson and strategic locations in Anchorage. Supplying water continued until April 5, when Anchorage was able to get its own system going again.

Some people lost their homes. Others lost power and could not heat their homes. Many were moved to shelters. In all, the Alaskan Command provided 244,000 gallons of diesel fuel and donated 800 sleeping bags, 200 bunks, 450 air mattresses, 80 regular mattresses, hundreds of electric blankets, 5,000 paper cups and plates, 2,900 feet of electric wire, and 2,240 pounds of chlorinated lime.

Fort Richardson and Elmendorf Air Force Base in Anchorage, the Kodiak Naval Station in Kodiak, and Fort Wainwright in Fairbanks provided critical volunteer labor and materials when people were in their greatest need. Soldiers from Fort Richardson remained on guard at Turnagain Heights until pulled back to base on May 1. That ended Operation Helping Hand.

Rushing requests for financial aid before Congress, adhering to the Bob Atwood attitude of optimism for the future, Anchorage and Alaska attacked the problems of cleanup, mourning, and rebuilding with an extraordinary passion.

14
Other Big Ones

The Alaska Experience Theatre is located in downtown Anchorage on Fourth Avenue. For more about thirty years, the exhibit, highlighted by a movie, has been telling the story of the Great Alaska Earthquake of 1964.

As part of the fifteen-minute show and narration, the darkened theatre shakes to recreate the feeling of being caught in a violent earthquake. The theatre's brochure says "It's a Seismic Sensation! Feel the ground actually shake and rumble beneath your seat . . . " The "Earthquake Simulator," it is explained, makes the seats "move during the show. So do not stand up during the show."

Twice during each show the theatre vibrates to shake up the audience. On Nov. 3, 2002, a couple emerged from the theater remarking how realistic it was in simulating a tumultuous earthquake experience. Especially, they said, the three times the theatre shook.

It so happened that the couple was attending the show about the 1964 earthquake when the 2002 Denali Fault earthquake struck. The show was so realistic because one of the three shakes was the real thing, a 7.9-magnitude earthquake.

That year, the 7.9 shake was the biggest earthquake anyplace in the world. Once again, Alaska was lucky because the worst-hit areas were remote. In this earthquake, the Denali Fault, on the west side in the Bering Sea, and the Wrangell Subplate collided. A twenty-four-mile section of earth was propelled upwards by thirteen feet, causing massive ruptures at ground level. Crevasses and cracks appeared on the Alaska Highway near Tok and shook parts of the trans-Alaska pipeline.

The oil pipeline that had been built to withstand the violence of a huge earthquake did hold up, even though the nearby Richardson Highway was damaged. However, no one died and there was no significant tidal wave. People were tossed about when they tried to stand or walk. An elder in the village of Mentasta told *Alaska* magazine, "I was like a log in a raft."

As powerful as that earthquake was, the Good Friday Earthquake of 1964 was one hundred times stronger.

Local circumstances and proximity to the epicenter mean that earthquakes that register much lower than 9.2, or even than 7.9, can initiate much worse results, with thousands of deaths, dangerous tsunamis, and aftereffects that linger.

Another Alaska earthquake of about 5.6 in May of 1995 was felt more widely in Anchorage than the 7.9 quake because the epicenter was only twenty-five miles from the city. While not resulting in any damage, except for the occasional broken window or falling glass objects, the 5.6 quake made people in Anchorage very nervous.

"Certainly, we're closer every day to the next big earthquake," Alaska geologist Tom Miller told the *Anchorage Daily News,* "but it doesn't mean one is imminent. It's definitely a reminder that we live in a very seismically active part of the world."

That is not something most Alaskans need reminding about. They know more about earthquakes than residents almost anywhere else. For people in other parts of the United States, a major earthquake might constitute a major surprise.

One major earthquake in the Lower 48 that most Alaskans may not recall, partially because it occurred in a remote area, occurred on Aug. 17, 1959. This massive earthquake altered the Montana landscape near Yellowstone National Park.

As more than two hundred people camped along the banks of the Madison River in southwestern Montana, settling into cabins near the shores of Hebgen Lake, the earth came to life with a 7.5-magnitude earthquake. Most people in the vicinity were vacationing, intent on fishing for rainbow trout and probably planned to get up early, meaning that most were in bed at 11:37 p.m.

When the Madison Canyon Earthquake, or the Yellowstone Earthquake as it also was called, began rumbling through the area, some campers awoke, believing a bear was shaking their trailers and tent. Some roads became impassable because of tumbling boulders. Like survivors in Anchorage, for a time cut off from the outside world, many worried that a bomb had fallen and a war started. Aftershocks of 5.8 to 6.3 only exacerbated that concern.

At the center of the thirty- to forty-second tremor was Sheep Mountain, a nearby peak that poured eighty million tons of rock down its slope at a speed of 174 mph into an area a half-mile wide near and into the Madison River. The river was suddenly dammed (later the area became known as Quake Lake), and with nowhere to flow, the water overwhelmed the dam at man-made Hebgen Lake. As a wall of water rushed twenty feet over the enclosure, houses were swallowed, roads devoured, and campers killed. The water level of Hebgen Lake rose eight feet, yet the old dam held.

Some twenty-eight people died that night in the Montana campground, most of whose bodies were never found, and eight others perished in Idaho. About $11 million in damage was reported. At Yellowstone, the underground thermal features were disrupted, with some new geysers created, and Old Faithful was knocked off schedule. A rock slide blocked some roads inside the park, and power and telephone service were lost.

Eighteen thousand people were in Yellowstone Park that night, so the toll could have been much worse if the earthquake had hit a few miles to the east. The Old Faithful Inn had to be evacuated, but elsewhere some overnight visitors never even realized there had been an earthquake. The book *The Night the Mountain Fell: The Story of the Montana-Yellowstone Earthquake,* by Edmund Christopherson and Elwood Averill, reported that an anonymous visitor at Canyon Village left this note on his pillow for the maid: "An awfully rough bear stayed under my cabin last night. Had an awfully hard time sleeping. Better tell the night man to do something about it."

Communications were knocked out throughout much of Montana, Idaho, and Wyoming. Rescue crews included smoke jumpers who normally parachuted into remote areas to fight forest fires.

The Madison Canyon Earthquake is one of the most powerful to hit the United States. Of the fifteen largest in U.S. history, thirteen have been recorded in Alaska, one in Hawaii, and one in the "Cascadia Subduction Zone," an area that includes Washington, Oregon, British Columbia, and northern California. That quake occurred on January 26, 1700, sending a tsunami across the Pacific Ocean where it damaged villages in Japan. Although little publicized in modern times, this earthquake is considered to be of 9.0 magnitude, ranking it the second largest in the United States.

The best-known U.S. earthquake was the San Francisco Earthquake of 1906. That quake measured 7.8 and wreaked far more damage and havoc than other large U.S. quakes because it struck in a densely populated area and caused immense fires that practically destroyed the city.

The San Francisco Earthquake struck at 5:12 a.m. on April 18, 1906. It is one of the nation's worst urban calamities. Its impact has been compared to the catastrophe of war, partially due to the three days of unchecked fire that rampaged through the city. At the time, the local fire department consisted of steam engines and horse-drawn wagons, and most of the city's wooden buildings had been constructed in the 1800s.

An estimated three thousand people died and $524 million worth of property damage was calculated. The closer a structure or road was to the San Andreas Fault line, the worse the damage was. Water mains ruptured, proving disastrous when fires spread. In Santa Rosa, about twenty miles to the north, fifty people were killed.

The earthquake's shaking lasted about one minute, during which a steamship headed to port was tossed about. In the book *The Great Earthquake and Firestorms of 1906*, author Philip L. Fradkin published this quote from the captain of the *National City*: "The ship seemed to jump clear out of the water, the engines raced fearfully, as though the shaft or wheel had gone, and then a violent trembling fore and aft and sideways, reminding me of running full speed against a wall of ice."

The earthquake collapsed so many buildings and fire consumed so many more that the event left sixty thousand homeless refugees.

Bizarrely, once San Francisco was habitable again and efforts to rebuild were swiftly put into place, ghoulish entrepreneurs made money off the tourist trade, advertising "See San Francisco in Ruins" or "On the Trail of the Greatest Fire in the World's History." For $1.50, one could take a two-hour car ride through "the most interesting parts of the burned district."

The next most-famous continental U.S. earthquake is probably the 1989 Loma Prieta Earthquake near San Francisco. That 6.9-magnitude earthquake interrupted the World Series between the San Francisco Giants and Oakland A's, and due to those circumstances, the earthquake actually was televised live.

Cameras were in place to telecast the third game of the World Series at Candlestick Park on October 17 when the earthquake hit, disrupting pre-game warm-ups at 5:04 p.m., just minutes before the scheduled 5:35 p.m. scheduled start. Sportswriters were in their places in the press box and were instantly converted into earthquake reporters. Many of the expected sixty-two thousand spectators were already in their seats when the shaking started. Players rushed toward their families in the stands, and many people quickly abandoned the grandstands for the open space of the field. The quake, centered in Forest of Nisene Marks State Park, caused the collapse of part of the Bay Bridge connecting San Francisco Bay's two largest cities, killed sixty-three people, injured 3,757, and left between 8,000 and 12,000 people homeless.

At the time, Oakland was leading the best-of-seven World Series 2-0. The earthquake provoked a ten-day delay in the middle of the series, stretching it from October 14 to October 28. The A's ended up sweeping the series, 4-0.

As in the Good Friday Earthquake when many had the day off from school and work, the fact that the World Series riveted Bay-Area residents took them off the roads and kept them at home in preparation for watching the game on television. Rush hour that day was less frenzied than normal.

ABC's Al Michaels, Jim Palmer and Tim McCarver, the baseball broadcast crew, began broadcasting footage of the earthquake, with Michaels as the lead reporter, especially after the network cut to Ted Koppel anchoring news coverage in Washington, D.C. The NBC crew on the scene—Jack Buck, Johnny Bench, and John Rooney—ran from its booth, whereupon Buck teased Hall of Fame catcher Bench with the comment, "If you would have moved that fast when you played, you wouldn't have hit into so many double plays."

When veteran Detroit sportswriter Joe Falls felt the tremor, he wasn't sure what it was. "I thought it was the usual pre-game flyover by some jets and I looked up to see the planes," he wrote. "I didn't see any planes. I looked down at the seats along the first base line into right field and everyone seemed to be holding onto something. The next thing I knew, I had jumped out of my seat and wound up on all fours on the concrete landing behind my row. I could envision the stands collapsing and everyone in the upper deck falling into the people in the lower deck. I had nothing to hold onto, so I just rocked back and forth as the concrete under me rocked back and forth."

And then Falls made a run for it, to get out of Candlestick Park, though he and others around him fleeing did something inexplicable. "Don't ask me to explain this, but when we got to the gate we all stopped to get our hands stamped so we could get back in."

They needn't have worried. No one went back in for ten days.

Oakland pitcher Dave Stewart, named Most Valuable Player in the series, recounted his emotions for the World Series program twenty years later. He said the scene contained "a sound that you never forget. It's the sound of tragedy. You can hear that a thousand miles away. You saw people crying. The faces—you never forget that."

ONE OF THE WORLD'S largest and most formidable earthquakes ever is still fresh in memory.

On Dec. 26, 2004, an earthquake centered in the Indian Ocean off the west coast of the island of Sumatra, Indonesia, registered 9.1 in magnitude, just behind the Alaska earthquake. Actually, various measurements place the quake at between 8.8 and 9.3, with the U.S. Geological Survey calling it 9.1.

The quake's duration stunned the scientific community, lasting between eight and ten minutes, the longest ever recorded for a massive quake. The combination of the supercharged quake and tidal waves that rushed across the ocean resulted in the deaths of 230,000 people in fourteen countries. Hardest hit were Indonesia, Sri Lanka, India, and Thailand.

National Geographic magazine cites the U.S. Geological Survey's analysis that the quake unleashed the equivalent of "23,000 Hiroshima-type atomic bombs." The magazine also suggested that with 150,000 people dead or missing simply in the aftermath of tidal waves, the tsunami may have been the deadliest in world history.

The earthquake's immense power when the India Plate collided with the Burma Plate caused the earth to move on its axis by one centimeter.

People running for their lives were washed off beaches. People unawares were swept away in houses. Vacation resorts, filled with Christmas tourist travelers seeking fun in the sun, were inundated. Vivid video from handheld cameras ricocheted around the planet on news shows and the Internet. Some of the tidal waves reached ninety-eight feet and virtually obliterated all life along long stretches of shoreline. Tsunamis traveled as far as three thousand miles from Indonesia to Africa. In many areas the tsunami arrived so quickly that it caught people unawares.

Dramatically, before psyches could heal from the huge catastrophe, a second earthquake measured at 6.3 struck Indonesia's Java Island, a densely populated area, in May of 2006, and killed more than 3,500 people. Homes and buildings not built to withstand strong tremors collapsed into rubble, leaving thousands homeless and sleeping in rice fields.

Then, on September 30, 2009, a third quake, measuring 7.6, struck Padang, Indonesia. Meanwhile, within a day, an 8.1 earthquake struck in Samoa, killing 189 people in Samoa, American Samoa, and Tonga. This was the largest earthquake of 2009. President Barack Obama declared a state of emergency in the territory of American Samoa, home of sixty-five thousand people.

Closer to the United States, Haiti was slammed by an earthquake on January 12, 2010 with colossal repercussions for the poorest nation in the Western Hemisphere. The virtually destitute island nation was ripped by a 7.0 quake that took a tremendous toll on a land where many thousands lived in shacks and shanties and where building codes were inadequate or nonexistent. The death toll exceeded 200,000; more than 300,000 others were injured.

The earthquake was triggered by the shifting of the Caribbean and North American Plates. The epicenter was ten miles from Port-au-Prince, where two million people resided. To compound the likelihood of worst-case disaster, the quake was shallow, about six miles below the ground.

Despite international aid, the ripple effects included cholera and people living amid rubble, in tents, and in the open for months. More than a million people were left homeless in a nation of nine million. The country has yet to recover.

The world's largest earthquake registering 9.5 was recorded on May 22, 1960, in Valdivia, Chile. That undersea earthquake killed at least 1,655 people, by one estimate, and possibly as many as 6,000. Another 3,000 were injured, while two million people were left homeless, and property damage was estimated at $550 million.

Although there was no technology to predict a super-quake like this, foreshadowing came in the way of several earthquakes, later dubbed "forequakes," registering between 7.0 and 7.7.

The 9.5 big one caused a tidal wave that roared across the ocean, killing 61 people and causing $75 million worth in damage in Hawaii, mostly in Hilo, hit by a 35-foot wave. Waves also killed 138 people and caused $50 million worth of damage in Japan and killed 32 people in the Philippines. The tsunami did reach Alaska, touching the Aleutian Islands. In Chile the quake also triggered volcanic eruptions, flooding, and landslides.

On February 27, 2010, a few days before a new president of Chile was to be sworn in, an earthquake with an 8.8 magnitude rocked the country. The Maule Region quake was located seventy miles offshore and immediately provoked fears of a tsunami. In a country that learned its lesson from the 1960 earthquake, with stiffer building codes and earthquake-preparedness classes given in elementary schools, the death toll was reported at 562—fewer than expected—though one million buildings were damaged.

Tsunami warnings were declared in fifty-three countries. If there was one area where Chile fell short in readiness, it was a lack of recognition of the danger of a tsunami to its own people. The government stated at first that the nation did not have anything to fear from tsunamis. That was incorrect. Forty retirees riding on a bus were swept to their deaths by a tsunami in Pelluhue.

Like Alaskans in the aftermath of the Good Friday Earthquake, Chileans relied heavily on the intervention of fourteen thousand soldiers to keep order, organize rescuers, and prevent looting. Vandals had been roaming the streets of Concepcion, the nation's second-largest city, and setting stores afire after looting them. Some arrests of looters made publicly at gunpoint were applauded by onlookers.

A few days later as a new president was sworn into office, the ceremony was interrupted by a 7.2 aftershock.

IN WHAT SEEMED like a constant series of major catastrophes, the world was reminded again how devastating earthquakes and tsunamis can be with the March 11, 2011, monster quake that struck Japan. The 9.0 magnitude quake in the Pacific Ocean slammed into the island nation with incredible force and produced a twenty-three-foot tsunami wave that swept away hundreds of people, spawned fires, and threatened a nuclear power plant. Later it was reported that a much larger wave, 133 feet high, had been sighted.

It was a triple whammy of violent earth shaking for two minutes, walls of water, and the fear stemming from a partial meltdown of a nuclear power plant reactor. Leaks were not brought under control for days as workers exposed themselves to dangerous levels of radiation in hope of averting an even larger disaster.

The threat from the nuclear reactor alone caused the evacuation of 170,000 people living in a twelve-mile radius, even as the country mourned its 21,000 dead or missing and sought to cope with the mounting catastrophe on several fronts.

Aftershocks exceeding 6.0 complicated matters after the initial thrust of the quake originating in the ocean about 110 miles east of Tokyo. Thousands of people vanished with the onslaught of tidal waves. The overall damage was immense, despite Japan being one of the most earthquake-savvy nations on earth.

More than 2.5 million homes lost power almost immediately, and 215,000 people were relocated to shelters and at first survived on biscuits and rice balls. About 5,700 people were injured.

The prime minister said it was Japan's worst crisis since World War II.

The quake was initially announced as a 9.0 magnitude shaking, and then listed as 8.9, still one of the five strongest earthquakes since 1900. Yet the earthquake was so powerful that the entire country shifted eight feet closer to North America and the entire planet Earth shifted four to ten inches on its axis as a result of the upheaval, according to NASA.

A tidal wave generated by the Japanese quake slammed into Hawaii where it sank or damaged two hundred boats. A warning was issued for Dutch Harbor in the Aleutian Islands, but no damage occurred there.

However, one of four California ports that did suffer tidal wave damage was Crescent City in an eerie echo of the Good Friday Earthquake. Three men were carried out to sea; two swam back safely. The third, who had been taking pictures too close to the water, was lost.

15

New Valdez

Remnants and reminders remain.

Just off the Richardson Highway there is a memorial to those who died in the Good Friday Earthquake and tsunami of 1964. Nearby are a few rusted-out hulks of shipping containers or remnants of boats, reminders of the original Valdez that existed in this location from 1898 to 1967.

The Valdez that people knew at the southern end of the Richardson Highway for three-quarters of a century had been built on an unstable base of silty, sandy soil, and it took the earthquake to reveal how unsafe it was. Even structures that remained standing were on shaky, cracked ground.

After the earthquake swept away the town's piers and claimed more than thirty lives, an investigation was launched to determine the suitability of rebuilding the pleasant little community facing the sea. Residents were shocked when the U.S. Army Corps of Engineers announced that the old town site would be condemned.

On October 1, services including electricity, telephone, sewer and water would be shut off. An instruction booklet was given to all residents, describing choices and options for moving. What was stressed in large letters was that everyone had to be out by September 30, 1967. Buildings that were not moved would be demolished.

Partially replacing old Valdez would be a 250-lot subdivision at Mineral Creek, four miles away. It was heralded as Alaska's first planned city. The new city included new streets, schools, hospital, churches, government buildings, a small-boat harbor, and city dock.

Residents were given two years to formulate and carry out a plan to move the community four miles away to the safer location and invent a new Valdez. People could build anew or move their homes and buildings to the new town site. But they couldn't stay where they were.

Concurrent with the studies of the physical landscape was an analysis of the earthquake situation. When it was determined that the repeat of a significant earthquake was likely to occur in a nearby location within fifteen years, that was the clinching argument for moving the entire town.

The decision may have been traumatizing, but it made sense. No one wanted a repeat of the horror wreaked by the Great Alaska Earthquake, and arguing to stay was pointless when everyone else was going. You could take your neighborhood with you.

At the time of the earthquake, Valdez had seven churches, a grade school and a high school, a state mental hospital, a local hospital that was characterized in one study as being more or less an outpatient clinic, two grocery stores, a clothing store, and a movie theater open four nights a week.

The economy was based on fishing, canning, shipping, government work, and a small level of tourism.

A report by James M. Tanaka for the Army Corps of Engineers cited past examples of earthquake-damaged cities being relocated. He referred to the Yugoslavian community of Scupi, hit by an earthquake in 518 that re-emerged as Skopje a few miles south. Guatemala Antigua became Guatemala City after an earthquake. However, Tanaka noted, officials had little information then about the likelihood of future earthquakes and did not relocate the cities far enough away from fault lines, leaving both to the mercy of future quakes. Valdez, he noted, was being moved with more deliberate thought.

On April 27, 1964, a month after the earthquake, the Valdez City Council adopted a resolution to move the community four miles west, still abutting the Port of Valdez, but on more stable ground.

Valdez became a city in transition. Planning and design for a new community were put into motion. By February 1967, the new community was in place. About fifty buildings were jacked up and moved. Some 750 people moved themselves. The cost was roughly $37 million.

For some people who evacuated in a hurry and had no home to return to, exile from Valdez lasted far longer than the day or two they originally imagined. Fred Christoffersen, whose family thought he was lost at the pier, said they all stayed in Glennallen for a year.

"When they rebuilt Valdez all of the power lines and sewer lines were broken," he said. "They had to be repaired and all of the tank-farm facilities had to be replaced because they caught on fire."

Tom McAlister did some of that work in the summer after the earthquake. His wife had left for Snohomish, Washington, after the earthquake, and he rented an apartment while working construction.

At the time, the McAlisters gave serious consideration to leaving Valdez permanently and came close to doing so. Tom had the engine of his old truck overhauled and bought a four-wheeler to pull behind it and carry the family stuff. The idea was to return to his roots in Washington State. But as the summer slipped into mid-September and McAlister was still working, his wife asked him, "You don't want to leave, do you?" He answered, "No, do you?" When she said no, they decided they were staying with Valdez.

Another half-century later, they are still there.

"We still have earthquakes all of the time," McAlister said. "There's been some that have moved things around pretty good. You wait for it to stop. I spent twenty-seven years with the fire service here in town between being paid and as a volunteer and I was in charge of emergency preparedness for a number of years. I know that as long as the old house stands up, I'm good. My wife and I can sit here for thirty days with no outside anything. We've got water. We've got food. We've got everything. I've even got a generator in the garage if we should need it to keep the furnace running. We've got a wood stove if we need that, and wood. We're prepared."

The way McAlister remembers it, when the Corps of Engineers drilled at the old site, it found nothing but silt as deep as 129 feet. McAlister's replacement house rests on bedrock.

As all of the talk about moving away or moving to a new Valdez swirled around her, Gloria Day felt the decision was pretty much made for her and her fisherman husband.

"I wouldn't know where to go," she said.

They stayed in old town as long as they could, but when the new Valdez came into being, they were able to move into state-subsidized housing and stayed in the community for good.

People who lived through the quake learned to live with earthquakes as a periodic nuisance that comes with the territory, but all of these decades later, this is a psychological residue left over from 1964 when a new quake hits.

"We get tremors out here every once in a while," said Thelma Barnum, who owned a gift shop in the 1970s in the Sheffield House that replaced the Wyatt Inn. "I can't do much. I think, 'Is it going to get worse?' because it starts slow and gets a little harder and then it usually goes away."

Barnum tries not to dwell on March 27, 1964. "I don't like to think about it," she said. "It brings back so much terror." Over time, she came to conclude that a "big one" like 1964 was not going to happen again in her lifetime.

NEW VALDEZ built a museum, where the earthquake is often discussed and on display. For a time Barnum worked in the museum, but it made her uncomfortable.

"I always felt a little guilty," she said, "because some of the people who lost their husbands or fathers on the dock came down to see what it was like. I feel so bad for them, you know, for their loss, teenagers, kids that were becoming teenagers, and they lost their fathers. Some of them I wondered what they'd be like if their fathers had lived. They're all good people now, but for a while some of them were so lost, I think."

Valdez did not forget the people who were killed by the earthquake and tsunami. On the first anniversary of their deaths, the community conducted a memorial service at the site of the dock that was destroyed. Later, the names of those who died were placed on a plaque at the old city site. Religious and civic leaders have commemorated the earthquake's anniversary over the years.

JOHN KELSEY, the last surviving member of the quartet of businessmen that owned the Valdez dock operation at the time of the earthquake, attributed the city's survival to the outlook of the people.

"If I were asked why we are where we are today, it is because of the frontier attitude of the people that stayed," Kelsey said in an interview years later. "Now there were a lot of folks that left Valdez and went elsewhere... There was a core area of people that stayed here and faced up to the challenges... I think they like challenges and that's one of the reasons they were here."

Valdez could have become a ghost town. Once-thriving mining towns elsewhere in Alaska did fade away.

"It would have been easy to abandon Valdez," Kelsey said, "and it was the choice of some people. And for many reasons, whether it was loss of jobs or loss of business, or their desire to get the hell out of Dodge. You know where the earthquakes are, and it was a frontier situation that could have tilted either way. Those of us that stayed, you know, I think we prevailed pretty well. I have an admiration for the people that did stay, I really do."

The new Valdez might have remained a sleepy town tucked in its own comfortable corner of Alaska, but it was chosen as the location for the southern terminus of the trans-Alaska pipeline. Despite its 306 inches of snow annually, Valdez now has more than 4,300 people, many of whom work for Alyeska Pipeline Service Company.

Since 1977, oil has flowed some eight hundred miles from Deadhorse on Alaska's North Slope to Valdez, where it is loaded onto tankers. Valdez was chosen for the pipeline because it was the closest ice-free port. Although tourism is more prevalent than it was in 1964, it is oil that makes Valdez tick these days.

LIFE WAS NEVER quite the same for the Bill Egan family after the Valdez earthquake. The governor and his family were living in the chief executive's mansion in Juneau when the earthquake struck. Like so many others, Egan lost precious possessions, although he didn't complain about it or focus his attention on it with so much other responsibility on his plate.

Even after the time period when Egan worked out of an emergency headquarters in Anchorage, toured Valdez, and returned to Juneau, Dennis said his father never talked much about his old home and how damaged it was. And Dennis didn't ask.

"He was very busy," Dennis Egan said years later. "He was a busy person and I knew better as a kid not to interrupt anything. I got all of the news from my mother, Neva. It was a horrible thing and my parents, you know, didn't talk much to me. They solved their own problems and didn't bother me. They knew there was no way in hell I was going to be able to solve them."

Neva Egan, who was from Valdez like her husband, died in early 2011 at age ninety-six.

Dennis said the family moved to Juneau when he was in the fifth grade. He finished school there and stayed to pursue a career in broadcasting. He even served as mayor in 2013 and was a member of the Alaska State Senate.

In the early 1970s, Dennis visited Valdez, noting that his uncle, Truck Egan, had moved his house to the new Valdez.

"Everybody was mad as hell that he moved his junky old house over there," Dennis Egan said recently. "And the damned thing is still there. The house is now a liquor store."

Although he still has many friends in Valdez, Dennis said he did not linger long on his first visit back after the earthquake. "It didn't sit very well with me," he said.

"I drove down by the old house. In fact, during pipeline construction they stored pipe where our old house used to be."

Bill Egan served as governor between 1959 and 1966 and again from 1970 to 1974. He died in 1984 from lung cancer at age sixty-nine. Anchorage's downtown civic and convention center later was named after him. It was built on Fifth Avenue, one block from where the Good Friday Earthquake of 1964 ruptured the streets of Alaska's largest city.

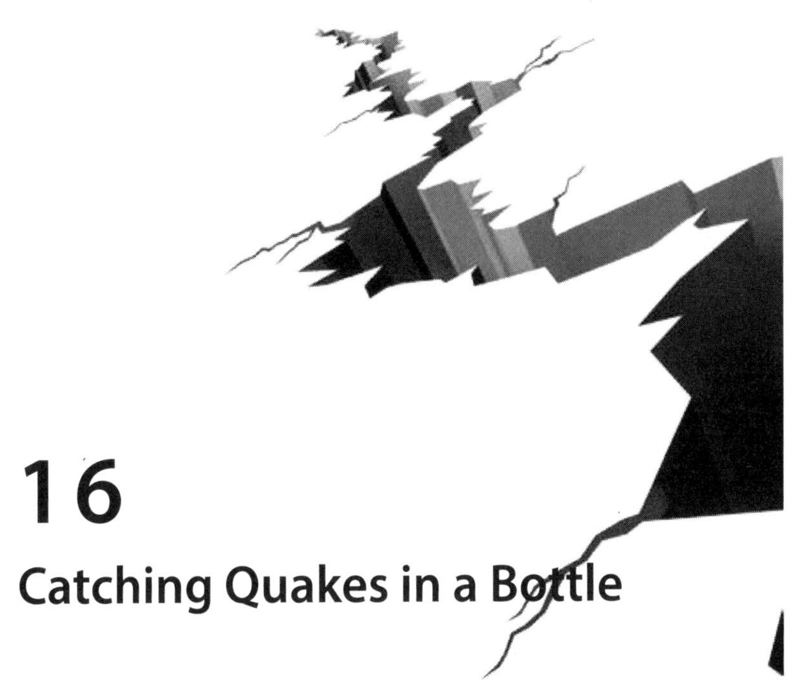

16

Catching Quakes in a Bottle

There is always some kind of earth-science work taking place at the Alaska Geophysical Institute on the campus of the University of Alaska Fairbanks.

Sophisticated monitors now chart the world of earthquakes, taking notice of even the tiniest shakes around the world. On any given day, dozens of earthquakes are recorded, even if they barely produce noticeable earth motion.

The machines actually talk out loud to their human observers: "There has been a 3.4 earthquake in Japan."

The West Coast and Alaska Tsunami Warning Center is based in Palmer, about forty miles north of Anchorage.

These Alaska-based scientific centers study earthquakes and monitor their deadliest effects, as humans continuously seek to understand and predict nature's forces, even if they cannot control them. The challenge of forecasting earthquakes remains an unconquered one. The best that can be done is discussion of the probability of an impending earthquake. But no one offers guarantees about a precise time when it might strike.

Tidal wave analysis is more reactive. Once a danger is recognized, speedy response can produce warnings that save lives, though how much in advance of a tsunami's strike such an alarm can be sounded is always iffy.

Nearly a half-century after the Good Friday Earthquake of 1964, nature is still running ahead of science.

Long-term preparedness, including more demanding building codes and training on how to respond when an earthquake or tsunami hit, remain among the best tools those studying earthquakes and resultant tsunamis can recommend to the public.

While it is easier than ever to predict the arrival of a tidal wave that might not strike land for eighteen hours, a beach community located much closer to the rolling waves might have less time to respond. At one point, scientists issued standing advice to anyone who felt the ground shake for thirty seconds or more: run inland as fast as you can go.

It may have been stated tongue in cheek when casually mentioned by a tsunami expert in 2001, but he was deadly serious in meaning when he said, "Don't go down to the beach to watch."

In 2007, scientists at the Geophysical Institute began a new study of all earthquakes 3.5 magnitude or higher and concluded that downtown Anchorage—one of the most damaged places during the 1964 earthquake—was at high risk for a harder hit than surrounding Chugach Mountain areas in the event of another big earthquake. Anchorage rebuilt in the downtown areas destroyed, sometimes using better technology to construct stronger buildings, though not in every case.

A highly motivated determination to rebuild, led by politicians and the *Anchorage Times*, existed in the days immediately after the Good Friday Earthquake. Exclusive housing in Turnagain, which had been shattered, was indeed rebuilt in the area. It was not completely unknown in 1964 that the houses overlooking Cook Inlet would be somewhat at risk in case of a large earthquake, though no one anticipated the scope of the calamity.

Still, on the very day of the earthquake, Mayor George Sharrock happened to be talking with banker Bob Baker, a Turnagain resident, over a drink at the Anchorage International Airport bar after dropping a friend off for a flight. Reportedly Sharrock joked with Baker about the precarious nature of owning a home in Turnagain, saying, "Bob, you'll wake up one of these days and you'll find your home in the Inlet."

Studies of the stability of the ground in Anchorage were made immediately after the earthquake. The soil was considered so iffy for future construction and expansion. Sharrock said that the report came close to recommending complete relocation of Anchorage. Alaska leaders did not consider that practical and ignored the comments about the lack of soundness for rebuilding in certain areas.

Like Bob Atwood, publisher of the *Times,* Wally Hickel was a made-in-Alaska power broker. A one-time Golden Gloves boxer with a pugnacious attitude, Hickel said in his autobiography that he arrived in Alaska with only pocket change and became a millionaire before twice winning election as governor. During this post-quake time of uncertainty, Hickel, grateful to the community that transformed his life, committed to building a high-rise luxury hotel downtown.

Hickel, who has since passed away, hired engineers to make his own studies and concluded that the land nearby shook, but the site of the hotel did not.

"You can't run from natural disasters," Hickel said in 1989.

From that plan rose the Hotel Captain Cook, one of the anchors of downtown Anchorage today. It is located a short distance from some of the land most violently shaken by the 9.2 quake in 1964 with no guarantee that it won't happen again.

The trans-Alaska pipeline was built to withstand earthquakes. Engineered to specifications that theoretically indicated it would withstand a super-quake like the Good Friday Earthquake, the pipeline also included a fail-safe system that would automatically halt the flow of oil in a strong earthquake. When a 7.9 quake rocked Alaska's Interior in 2002, the pipeline's self-monitoring system did just that.

Dr. Gary Carver of Kodiak was part of the team that studied earthquake fault hazards in 1972 and 1973 before the pipeline was built. The work was done because the pipeline would cross the Denali Fault in central Alaska.

Carver said, "Our job was to look at the entire length of the pipeline, the proposed route of the pipeline where they ultimately built it, and find the active faults and characterize them so we could provide engineers with the information they needed to make the design and build the pipeline so it would not fail in the event one of these faults moved."

The assignment was to provide "fault parameters" for the engineers as precisely as possible that would determine how much vertical or horizontal motion would occur in the case of a major upheaval.

"We could narrow it down to this fairly short section of the pipe where the fault would be somewhere underneath that," Carver said.

When the November 3, 2002, quake struck, Alyeska crews inspected the pipeline, which runs both underground and above ground, and found some damage to the infrastructure between miles 588 and 589. Brackets that hold up the pipeline cylinders, which measure forty-eight inches, were damaged in eight places. That area, about fifty miles south of Delta Junction, was closest to the Denali Fault line, source of the quake. In addition, two supports broke, but the pipeline did not sag to the ground.

During that quake, Carver said, "The fault moved laterally eighteen inches. They had designed the pipe to accommodate that much motion."

That is why the pipeline was built above ground in that region. Huge "I beams" served as "sliders," Carver said, so the pipe could move with the ground when a quake hit.

"It worked," he said.

After the early 1970s research project was completed, Carver moved back to the Lower 48 and taught for twenty-five years. He is a paleoseismologist, someone who studies prehistoric earthquakes with regard to assessing fault lines to help understand the present and future. In 1998, he settled in Kodiak, where he found that the average person who had lived long enough still frequently had his mind on the tsunami from the Good Friday Earthquake that destroyed the city.

"The tsunami here is locked in everyone's minds totally," he said.

Carver said that the knowledge gained from the Good Friday Earthquake is what scientists hope will lead to more accurate predictions of future quakes.

"Predicting earthquakes is elusive, and at this point to make a specific prediction where you give a specific time and characterize the event at that particular time is still something over the horizon," he said. "We don't have a way of doing that, but we can make forecasts much like the weather bureau makes forecasts, that there's going to be a wet winter, or there's going to be a storm coming. It's better than not knowing anything, especially if you're involved, for example, in building a nuclear power plant, or a very large, expensive pipeline, or the Bay Bridge that is going to have a one-hundred- or two-hundred-year life. Then the forecasting approach, this earthquake characterization of fault activity, becomes a very useful thing."

More is known now about the risk of tidal waves for Kodiak than was known in 1964.

"There's a difference between science and then the application," Carver said. "Sometimes the science is ahead of the application. If we had a 1964 earthquake right now in Kodiak it would do huge damage because the waterfront area is all built up again. We are one of the largest fishing ports in the nation, so we have fish-processing plants that line our entire shoreline in town.

"Here's the other thing we know. The likelihood of that earthquake repeating itself in Kodiak in the life of our fish-processing plants is virtually zero."

That's because the same scientific data indicates that another quake of such magnitude won't occur until between 330 and 700 years following the 1964 quake, Carver said. Unless a person is courting immortality, there is not a strong reason to expect being around for another quake of such significance.

MOST EXPERTS agree that a 9.2 earthquake now would be much more devastating in Alaska than it was in 1964 because of the rise in population. Anchorage had about three hundred thousand people in 2012, compared to sixty thousand in 1964. Valdez, Seward, and Kodiak have populations today that have grown about 400 percent in the past fifty years.

"Buildings are taller," said Mike Doogan, the former newspaper columnist. "Some of the sort of transportation infrastructure you've got now is probably more susceptible to damage. You've got overpasses and all the rest of that stuff. The likely effect of a similar earthquake would be more damaging, more costly, likely to basically kill more people. More casualties."

The enthusiasm for rebuilding Anchorage and other parts of Alaska gripped people demoralized by the earthquake and the tsunami. The energy of this effort impressed many.

Chuck Lastufka, the pharmaceutical representative who took to the air to deliver prescription drugs to patients soon after the quake hit, was good friends with some KLM Airline pilots who had flown all over the world and seen other earthquake devastation. They could not believe how swiftly Alaskans built, he said.

"Buildings started going up," Lastufka said. "Roads were paved. They (the pilots) were coming regularly and two years after the quake they were amazed at how Alaska had worked to rebuild. They had seen places around the world where there had been big earthquakes like in Iran or other places, and twenty years after they had an earthquake it looked as if it had happened yesterday."

Lastufka witnessed the fuel tanks burning in Seward. Given extensive damage from the earthquake, tidal wave, and fire, he wondered if the city would ever be rebuilt.

After the first shaky weeks, though, Dan Seavey and his wife decided against returning to Minnesota, and their lives gradually began to regain some normalcy. That summer was about families making personal assessments of their futures and Seward making its own self-examination.

"It got rebuilt and it was better in a lot of ways," Seavey said. "We talked a lot about going, but also about how the town was down and how could you leave then. I think the earthquake made us both dig in."

Long after Linda McRae MacSwain's family rode the waves of a tsunami on the roof of a house, the slightest earth tremor gets her attention.

"I'm really, really nervous about earthquakes," she said. "I have flashlights under everybody's bed. It's really awful. When I taught school in Fairbanks I told students: 'I had a really bad experience. You're supposed to get under your desk and this, that and the other, but I'm probably not going to be the best person to help you through this.'"

17

Anchorage Today

The park is pleasant and welcoming on a sunny summer afternoon. Bicycle riders pedal on a concrete path. Families with small children on foot share the same path that takes them into the trees. A mother moose and two calves halt traffic as they amble past. Earthquake Park in the Turnagain area of Anchorage commemorates the 1964 Good Friday Earthquake and serves as a memorial for those who died.

Plaques explain what happened on March 27, 1964:

"One of the most devastating earthquakes to hit North America struck Alaska …" "Centered in Prince William Sound, 80 miles south of Anchorage …"

"Greater force than the Mount St. Helens eruption…"

"100,000 square miles of the earth's surface was vertically and horizontally displaced …"

"North and west of the fault line land dropped 2 to 7 feet …"

"90 seconds into the Good Friday Earthquake an 8,000-foot strip of bluff, 1,200 feet wide, began cracking apart into large blocks which slid toward Cook Inlet …"

"Some homes slid 500 feet; others just broke or were crushed."

The lesson offered by photographs of houses split in half and tumbling down the bluff remains vivid for those who lived through the calamity. Over the years the clamor to rebuild on the same soft ground gradually outweighed common sense, and new homes were built in the slide area.

In 1985, Lidia Selkregg, a geologist, tried to inject reality into the discussion.

"The people in Turnagain are foolish," she said. "That's my opinion. We now have the ability to prevent another disaster. What kind of evidence do people need? Turnagain was destroyed."

The neighborhood has since risen from the debris, and there is no way of knowing whether homes would withstand a repeat of the violent shaking of 1964. A 9.2 earthquake might strike again soon, or a quake of such power might not occur for another century, when all of the optimists who moved back to Turnagain have passed away.

Although more attention is lavished on the likelihood of Los Angeles and San Francisco experiencing a major earthquake—"The Big One," as it is often spoken of—geologists agree that Alaska is much more seismically active and that Anchorage runs a larger risk of getting slammed than either California city.

For those who survived the terror of the 1994 Northridge, California, earthquake that killed 61 people and injured 9,000 while leaving 25,000 people homeless, the knowledge that it registered only 6.7 is sobering. Anchorage's 9.2 quake was 10,000 times more powerful.

The experience of those who lived through the 1964 Alaska earthquake in Anchorage and stayed in the city is seared into their memories. They have lived through countless earthquakes in the decades since, and most of those quakes made them nervous. Always they wondered if they were about to experience a once-in-a-lifetime earthquake for a second time.

Mike Doogan says the longer an earthquake lasts, the edgier he gets.

"The thing about an earthquake is that you never know how big it's going to be or when it's going to stop," he said. "So I think that if you were around when it happened (the Good Friday quake), you're a heck of a lot faster to the door jamb than you might be otherwise."

In 1964, it took more than a week to regain some normalcy for school children in Anchorage. Kids stayed at home with their parents while the city sorted itself out. One of the seriously damaged buildings was West High School, located near Turnagain. The second floor collapsed onto the ground floor. It is frightening to contemplate what the tragedy would have been like if school had been in session.

"Massive casualties," said Mike Janecek, who knows that Anchorage got off easy. "There were so many aspects. Within the first hour we had the earthquake. We're all terrified. In fifteen minutes people were talking about the potential of a tsunami coming. That scared me. They were talking about hundreds of feet of water. So that was terrifying. Then came the 'Oh, the gas main is burst down at the port.' There was a guy that realized what had happened and went down to the port area and turned off the gas. Otherwise it could have been a total disaster area."

Janecek said the big quake scarred him.

"To this day, I react differently than most people do to an earthquake," he said. "I'm very sensitive. If there's shaking, I want out. If the school district that I work for said, 'you do the drop and get-under-the-desk drill,' I'm leaving, and if that costs me my job, I'm gone. I'm telling those kids, 'It's every man for himself. We're out of here, kids.'"

Janecek has experienced earthquakes exceeding magnitude 7.0 since 1964, and "it stops my heart. I'm always up and going. I want out of a building. If I'm outside, I'll tough it out, but I want to go. I had to remain in the building one time. My mother was in the hospital on the fifth floor of Providence, and I couldn't leave her. I had to stay there and it was the most terrifying thing I ever did."

One of the distinguishing traits—and the most memorable—of the Alaska earthquake was its longevity, lasting more than five minutes. Smaller earthquakes don't last nearly that long, so when the ground begins shaking, Janecek is also conscious of how long a jolt lasts.

"After a couple of beats, I'm worried," he said.

To the point where he has flashbacks; never buried very deeply in his memory is the damage that afflicted the JC Penney store.

"Honestly, the image of JC Penney is still etched in my brain," Janecek said. "Dust coming up, the crushed cars in there. The instant realization that somebody died there, it stopped you in your tracks and turned the event into a whole different game. Suddenly, it wasn't an adventure, it was really scary."

Janecek, a retired high-school teacher and track coach who served Palmer High School as athletic director for ten years, has told his earthquake stories to young people many, many times. He said he does it because "it is part of our culture and it defines how we live. I want them to be vigilant. I want them to be safe about it. To this day I wonder why we built on some of the places again that we did.

"It never changes. It's like human nature. I say it's the pregnancy issue where something happens in your brain after pregnancy that allows you to forget the pain that was involved so you can do it again; same thing with earthquakes."

One thing Alaskans were grateful for was that there were not more deaths attributed to the Good Friday Earthquake. A smaller, more spread out population helped account for that fact.

"What separates 1964 Anchorage, or Alaska, southcentral Alaska, from Japan is that we had 115 deaths and they had thousands because of the population density," Janecek said. "One aspect that I champion is to raise good, smart engineers to deal with buildings (to protect against future quakes). It's not going to go away. It's not going to change. To ignore it and build buildings that fall in on people is crazy. So that's one reason to constantly remind people of the history and culture."

Clevey Cooper, the man who was in a bar on Fourth Avenue when the Good Friday Earthquake came to visit, took note of a 5.1 earthquake that struck the Anchorage area in 2009.

"We felt it pretty good," he said. "But it just rolled away; as long as it stops with no damage. There's a big difference between ten seconds and five minutes. If there had been any fires (in Anchorage in 1964), it would have been all over, but we were lucky we didn't have that earthquake the way they did in San Francisco (in 1906)."

A half-century later, Steve Nerland finds it no challenge to remember the day the family furniture store vibrated so powerfully.

"Those memories and those pictures still appear in my mind," he said. "It was pretty traumatic because of the unknown factor. Rumor central was significant. The tidal wave scares for one."

Being shuttled to Fairbanks after the quake to stay with grandparents, Nerland missed some of the immediate aftermath of Anchorage's trauma. But he was right in the heart of it on the day of the quake, and for him that was enough. Like many others who felt the strength of the quake, Nerland comes to full attention today anytime the ground starts shaking in Anchorage, no matter how lightly.

In Nerland's mind, Anchorage was fortunate.

"So few people were killed and injured, sort of by happenstance," he said. "Considering that the Fourth Avenue Theatre dropped … Our store is on Fifth and the buildings on Fourth fell. It's a powerful recollection that we looked outside and saw that, man."

For Rick Nerland, Steve's younger brother, the quake's shaking was probably more horrifying because he was out in the street watching that very street crack open. He felt very vulnerable.

"My biggest fear was that somehow the ground would open up and I would get swallowed in," he said.

At that time Nerland had a newspaper route, delivering the *Anchorage Times*. Although kids weren't allowed into the bars, sometimes one of the paperboys stuck his head in, trying to make a quick sale by shouting "newspaper!" The *Times*, he said, used to sell for fifteen cents, and it was a big deal when a generous tipper told you to keep the change from a quarter.

After the earthquake, he said, a story went around among the paperboys that one boy who sold papers downtown had stuck his head into a bar just as the earthquake hit and saw a man so drunk he had his head down on the bar. When the room began shaking, he lifted his head and shouted, "Go, Mother Nature! Go!"

Rick Nerland remembers the first day or so following the quake when Anchorage was in emergency mode. Messages were being broadcast over the radio to let loved ones know others were okay and drinking water came from Army trucks.

"Boy, we sure listened to the radio a lot," the younger Nerland said. "You would bring your own container to the schools to get fresh drinking water. Fortunately, there was still snow on the ground so you could scoop some snow up and boil it. That worked for most purposes other than drinking water."

Radio reports were a public service, informing people about where they could obtain water, warning them to stay away from downtown and Turnagain, and also serving as a comfort to those worried about missing relatives. The messages broadcast, in Nerland's memory, went something like this: "This is to Bob and Mary Jones. Your daughter says she's fine. She's at such and such a place and will stay there."

Even in 1964, as a kid, Nerland said it was not difficult to gauge the mood of survivors who surveyed the devastation and realize how fortunate the city really was.

"There was a collective sense of that it was bad, but it sure could have been worse," he said. "There could have been schools full of kids. It could have been in the middle of the night. More people would have been in those homes that were swallowed up in the Turnagain area. I remember having a sort of distinct feeling that not only was my family fortunate, but the whole community was sort of issuing a sigh of relief that it really could have been worse."

As an adult, with the experience under his belt, Nerland has been a kind of human earthquake monitor with a built-in knowledge of awareness that this quake is nothing or that quake is something to be concerned about.

"You know you can kind of tell when there's just a little shake and a little rumble," Nerland said. "But then if it sort of amps up a little bit … well, that feeling comes right back. It comes right back remembering the strength of that event and the power of that earthquake. We had a few little earthquakes before 1964 when I was growing up, but they were just novelties. I don't ever remember them in Fairbanks, but in Anchorage it was, 'Oh, that was a little earthquake. Oh, wow, interesting. A novelty.' After the big one there were continual aftershocks, so it was important to be able to kind of read it. This is a just-get-in-the-doorway earthquake, or this is a run-outside earthquake."

The Nerlands had been Alaskans for more than half a century in 1964. Jerry Nerland said the family gave very little consideration to abandoning the state. When the earthquake was described as a one-hundred-year geological phenomenon, he chose to believe that.

"We hoped that our luck would be better," Jerry Nerland said. "As long as there had been one, we figured there wouldn't be another one."

Dick Mackey stayed with the Alaska Railroad for a few months after the earthquake. Track between Anchorage and Seward was mangled, but trains continued to travel northward. For a time, like many others, Dick mostly helped to clean up the mess. Later, he became a full-time iron worker who helped repair and rebuild Anchorage buildings wrecked by the quake.

"In some cases they demolished them and put up new ones," Mackey said, "a couple of them immediately. Then I went to Kodiak—what a disaster that was—and then to Seward. I actually spent the next couple of years working in those two places. The disaster level was unbelievable, especially in Kodiak. At first our crew of iron workers didn't have any place to stay. We stayed in tents. The first job we did was on a hotel. When we got a room done two guys slept there. There were about eight of us—carpenters and all of the trades. It was quite an experience. You'd go to the bar at night. It was in a tent. Kodiak was devastated, boats up on the hillside and everything."

In Seward, Mackey helped build a new railroad roundhouse. "And they were dredging the harbor," he said, "so there was a tremendous amount of activity. I think Seward had a population of about one thousand residents at that time and there were six hundred construction workers on the scene. It was chaotic."

Until then, Mackey said, all construction in Alaska was seasonal, but there was an urgency to rebuild. Nobody wanted to wait around with men and equipment idle.

"We learned to build under Visqueen," Mackey said. "It changed the whole concept of construction in Alaska. For an up-and-coming dog musher it was a disaster." Winter used to be the time for running sled dogs. Instead, Mackey was working full-time all year. For a couple of years he concentrated on ironwork.

Mary Inez Burkhart's pot roast, scheduled for dinner on March 27, was never eaten. There was no other food in the house, and the Burkharts went to bed with empty stomachs that night. The Burkharts also did some scrambling in the first days after the earthquake. She and her husband fixed up their camper and made a toilet out of a cardboard box, and then obtained water from the military trucks.

"They put up a great big canvas water container with a ladder," Burkhart said, "and they put a guard on it. He stood there as you climbed the ladder and reached down and got some water."

The Burkharts drove around town trying to do some errands. They went to a bank, hoping to withdraw some money, but the bank was closed. They went to a grocery store in Anchorage's Spenard area, and pretty much the only things available were canned goods that had fallen from shelves and clogged aisles.

The fourplex the Burkharts moved into after the sale of their house and their trip to Hawaii was fine, but the buyers of their house never made another payment. The earthquake drove them away.

"They were scared to death," Burkhart said. "We had to move back into that house, but it was a good area and got no damage. It all could have been so much worse."

About a week after the earthquake, Burkhart made an appointment at a beauty shop. Her hair had frizzed in Hawaii; anyway, being without running water didn't do the hairdo any good, either. The beautician had just added bleach and color to her natural hair when the building began shaking from a major aftershock.

"I said, 'Rinse this immediately! I'm out of here!'" Burkhart said she shouted. "I left there with my hair halfway rinsed, I was so terrified. I had to get back and see if everything was okay at my house."

Some fifty years later, memories of the awful quake still linger beneath the surface of her thoughts.

"When it starts shaking the least bit," she said, "you can rest assured before it has time to really get going, I'm already in my car and trying to get out of the garage. It'll never leave you. You see some of those things so vivid in your mind, they will never leave."

Al Bramstedt, the future television station operator, said his family came up with a different solution to the shortage of drinking water.

"We had no stored water and my mother said that the only thing we could do was drink water out of the toilet, out of the water closet," Bramstedt said. He and his three sisters responded rather negatively to that suggestion. "I remember my three sisters and I and my mom in the bathroom looking at the toilet. My mother said, 'This water is fine. There's no problem with it.' All of us said in unison, "No way we're going to drink out of the toilet.' She said, 'No, you're not drinking out of the toilet down there.' She pointed to the bowl. 'You're drinking out of what they call a water closet.' That's where the water is stored. 'No way!' we yelled again. What we ended up doing is that the electricity came back on and I shoveled snow from the backyard through a bathroom window into the tub and we took that snow and melted it for drinking water."

Bramstedt has reflected many times over the years about how lucky his family was that no one got hurt, how lucky Anchorage was that more people didn't get hurt, and how such a huge earthquake could have done more damage and made things really rough for people if it had occurred in January and the temperature was minus 20.

"I've thought about that every time there was another earthquake," Bramstedt said. "Is this the beginning of another big one? When you're a parent it changes your perspective and when you own things that you're trying to protect, it's no longer an exciting adventure, mainly a matter of trying to assess the potential risk. We were really fortunate."

Normally Bramstedt is acutely aware of earthquakes, but a couple of years ago he was at his remote cabin at Echo Lake when he was fooled by one. He had had heart surgery with a valve replacement, and the doctor told him that if he was going to indulge in heavy labor, not to allow his heart rate to get too high.

He was chopping wood and hauling logs one day when an earthquake hit. He didn't realize it at first, blaming his body. "I've just gotten myself over-active," he said. Bramstedt sat down on a stump and talked to himself. "Boy, you need to sit down. You're working too hard. Your heart is starting to screw up here." He then returned to Anchorage and told his brother-in-law about the wood-chopping incident. He was informed, "You goofy son of a bitch. It was a damned earthquake."

For a man who was general manager of a television station for thirty years, Bramstedt's antenna was not working too well that day.

Teacher Sarah Burkholder suffered an unusual loss during the earthquake. Her ten-gallon fish tank had its own tidal wave from the hard shaking and tossed her goldfish onto the floor, where they died. She also knew people who fled Alaska soon after the earth stopped moving.

"One was a gal from my hometown," Burkholder said. "That had been her first year up here. She was teaching and didn't sign another contract. She went back to Buffalo and taught for a year and then ended up in Hawaii, where she still is. Another couple, they had been having marital problems and whether or not the earthquake was the kicker, I don't know, but they left. They're still together, but they've never been back to Alaska."

Burkholder is still sensitive to earthquakes, too.

"Even today, if my husband is sitting on the couch and I'm on the other end and he starts shaking his leg and the couch begins to wiggle, I think, 'Oh my gosh, is that a big one, or is it a small one?'" she said. "That sensation never leaves."

Given the scope of the catastrophe, there was not much humor to be gleaned from the earthquake's violence, nor in the ensuing days, but Peggy Bensen recalled a friend who had retrieved her car from a repair shop and was driving home on Northern Lights Boulevard. The car swayed this way and that, and she was furious, cussing the repairmen for messing up her car and knocking the steering out of alignment.

"She was so mad," Bensen said—until she realized the issue was not her car at all, but one of the biggest earthquakes of all time shaking the road.

Speed Gratiot was not one of the earthquake survivors who ever wondered how much worse things could have been because he saw enough to be convinced the Good Friday Earthquake was bad enough.

"I was so awestruck by the damage that was done," he said. "My whole attitude was that I couldn't believe the earth could have that much damage and still stay together."

Gratiot worked on repairing the Anchorage International Airport control tower and found it unreal how hard hit that structure was, and he spent time in Kodiak a few years later during a period when damaged pipes and wiring were being replaced at a power plant at the naval base.

"I could see that some things were still a mess," he said. "We were down there a couple of months rewiring. I was on the crew that pulled wire out of pipes and oil and salt water all came out. We had to cut pipes loose and jack the wire out. It was all lead-covered wire, 5,000-volt wire with lead covering on it. We'd get oil and salt water all over us. There was nothing you could do about it. We'd work half a day, take a shower, then work another half a day, go home and take another shower."

Young Karen Sobolesky, who had been out snowmobiling with her mother, heard some funky stories after the chaos calmed. There was the one about the man naked on Fourth Avenue who had been in a steambath. An artist friend, who had been painting a Volga River scene because her military father had just been there, ended up with a variety of tributaries because of her brush being jerked around. Then there was the case of the missing pie.

"One woman had two pies in her oven when the earthquake struck," Sobolesky said. "At the end of the earthquake she had one pie. One of her pies was nowhere to be found. It wasn't on the floor. It was just gone. A couple of days later she opened up the drawer next to her oven and there it was. At one point during the earthquake the door of the oven flew open at the same time the drawer flew open and one pie flew out and landed perfectly upright in the drawer."

Today, Alaska considers itself far more prepared for a major earthquake than it was in 1964. More buildings have been constructed according to stricter codes. The public has been given more instructions about what to do in an earthquake.

And then there are people like Mike Janecek, who will never forget what they experienced and are happy to tell anyone who asks what it was like during that frightening time in Anchorage. He is out of the classroom now, but invitations to talk about the earthquake still come, and he still accepts.

"I never forget how significantly etched it is in my brain," he said. "I can close my eyes and see exactly what Fourth Avenue looked like when it happened and I think when something that big happens to you that's what occurs. Maybe I revisited it a lot more than other people by virtue of telling it to other people, but it is etched in your brain forever. The tiniest little nuances are etched in your brain."

All who lived through the Great Alaska Earthquake agree that there was nothing tiny about it.

Author Sources

PERSONAL INTERVIEWS: Thelma Barnum, Peggy Bensen, Al Bramstedt Jr., Mary Inez Burkhart, Sara Burkholder, Gary Carver, Fred Christoffersen, Clevey Cooper, Gloria Day, Mike Doogan, Dennis Egan, Carolyn Floyd, Joe Floyd, Flip Foldager, Erldon Gradiot, Roger Hansen, Mort Henry, Mike Janecek, Chuck Lastufka, Dick Mackey, Rick Mackey, Tom McAlister, Linda McRae MacSwain, Linda Myers-Steele, Jerry Nerland, Rick Nerland, Steve Nerland, Rita Ramos, Raymie Redington, John Schandelmeier, Dan Seavey, Karen Sobelesky, Dana Stabenow

NEWSPAPERS: *Anchorage Daily News, Anchorage Times, Chicago Tribune, Columbus* (Indiana) *Republic, Fairbanks Daily News-Miner, Honolulu Advertiser, Humboldt Times, Ketchikan Daily News, Kodiak Daily Mirror, Los Angeles Times, San Francisco Chronicle, San Francisco Examiner, Seward Phoenix Log, Valdez Earthquake Bugle, Valdez Vanguard*

WIRE SERVICE: Associated Press

MAGAZINES: *Alaska, Kodiak Tourism, Life, Montana Quarterly, National Geographic, Popular Mechanics, Time*

BOOKS: *Alaska's Homegrown Governor,* Elizabeth A. Tower, 2003; *Bob Atwood's Alaska,* Bob Atwood, 2003; *Tales of Alaska's Bush Rat Governor,* Jay Hammond, 1994; *Earthquake!* Eloise Engle, 1966; Joe Falls, *50 Years of Sportswriting,* 1997; *Many Battles: The Autobiography of Ernest Gruening,* Ernest Gruening, 1973; *Richter's Scale,* Susan Elizabeth Hough, 2007; *The Day That Cries Forever,* John Smelcer, 2006; *The Night the Mountain Fell: The Story of the Montana-Yellowstone Earthquake,* Edmund Christopherson and Elwood Averill; *The*

Day Trees Bent to the Ground, Janet Boylan, 2004; *The Great Earthquake and Firestorms of 1906,* Philip L. Fradkin, 2005; *Seldovia Alaska,* Susan Woodward Springer, 2005; *The Strangest Town in Alaska: The History of Whittier, Alaska and the Portage Valley,* Alan Taylor, 2005; *Alaska Earthquake '64,* compiled by Joy Griffin, 1996

FILMS: *Alaska! Earthquake 1964, Between the Glacier and the Sea, Ocean Fury: Tsunamis in Alaska, Though the Earth Be Moved*

WEB SITES: Alaska Geophysical Institute, msnbc.com, West Coast and Alaska Tsunami Warning Center, Wikipedia—list of world's largest earthquakes, Yahoo.com

OTHER: "A Narrative of the Human Response to the Alaska Earthquake, Valdez, Alaska," Frank R.B. Norton and J. Eugene Haas, Disaster Research Center, The Ohio State University; Alaska Experience Theatre; Anchorage Fire Department 75 Years, 1915 -1990; Earthquake Park; Earthquake preparedness brochure, Municipality of Anchorage;

National Science Foundation/Office of Naval Research, "The Great Alaska Earthquake March 27, 1964," George Pararas-Carayannis; "Operation Helping Hand—the United States Army and the Alaskan Earthquake, 27 March to 7 May, 1964," Major General Ned D. Moore; Personal correspondence, Anchorage Loussac Library/Valdez Museum;

"Report of U.S. Corps of Army Engineers District Alaska, Assistance Provided by the Military Bases," Thomas L. Gardner and Edmund J. McMahon; "The Alaska Earthquake Effects on Communities," U.S. Geological Survey professional paper; "The Great Alaska Earthquake of 1964/Committee on the Alaska Earthquake of the Division of Earth Sciences National Research Council," National Academy of Sciences, Washington, D.C., 1973; World Series program/Major League Baseball 2009

Index

Boldface indicates photo content

About the Author

LEW FREEDMAN is a freelance writer and is the author of fifty-six books, many of them about Alaska, where he lived for seventeen years and experienced numerous earthquakes.

An award-winning journalist who graduated with a Bachelor's Degree from Boston University and a Master's Degree from Alaska Pacific University, Freedman lives with his wife, Debra, in Indiana. Among his titles for Epicenter Press are *Iditarod Classics, Fishing for a Laugh, Iditarod Dreams* with musher DeeDee Jonrowe, and *Yukon Quest*, the history of the 1,000-mile race between Whitehorse, Yukon Territory and Fairbanks, Alaska.

Epicenter Press
ALASKA BOOK ADVENTURES™
www.epicenterpress.com